034078

GOULD, PHILLIP L
DYNAMIC RESPONSE OF STRUCTURES
000340782

624.042 G69

DYNAMIC RESPONSE OF STRUCTURES
TO WIND AND EARTHQUAKE LOADING

Dedication

To my wife Deborah for her continued inspiration and support

Phillip Gould

Dedication

To my parents, in their homeland Palestine, for their dedication to knowledge and truth.

Salman Abu-Sitta

DYNAMIC RESPONSE OF STRUCTURES TO WIND AND EARTHQUAKE LOADING

Phillip L. Gould
*Professor of Civil Engineering,
Washington University, U.S.A.*

and

Salman H. Abu-Sitta,
Consulting Engineer, Safat, Kuwait

PENTECH PRESS
London : Plymouth

First published 1980
by Pentech Press Limited
Estover Road, Plymouth,
Devon

© Pentech Press Ltd, 1980
ISBN 0 7273 0403 8

British Library Cataloguing in Publication Data
Gould, P L
    Dynamic response of structures to wind and
earthquake loading.
    1. Earthquake engineering. 2. Wind-pressure
  I. Title
    624'.176   TA654.6

ISBN 0-7273-0403-8

Filmset by Mid-County Press
Merivale Road, London SW15
Printed and bound in Great Britain by
Biddles Limited,
Guildford, Surrey

Contents

1 **STRUCTURAL DYNAMICS OF SINGLE-DEGREE-OF-FREEDOM SYSTEMS** — 1
  1.1 General — 1
  1.2 Free Vibration — 3
  1.3 Periodic Forcing Function — 6
  1.4 Response to Impulse and Convolution — 10
  1.5 Base Disturbance Model — 12
  1.6 Elasto-Plastic Model — 13
  References — 14

2 **RANDOM PROCESSES** — 15
  2.1 General — 15
  2.2 Statistical Parameters — 16
  2.3 Probability — 16
  2.4 Multiple Random Variables — 19
  2.5 Autocorrelation and Power Spectra — 21
  2.6 Combined Random Processes — 25
  References — 27

3 **SDF SYSTEMS UNDER RANDOM LOADING** — 28
  3.1 Simple Oscillator — 28
  3.2 Response of SDF System to Wind — 31
  3.3 Multiple Loads — 33
  3.4 Extensions — 34
  References — 34

4 **STRUCTURAL DYNAMICS OF MDF SYSTEMS** — 36
  4.1 General — 36
  4.2 Equations of Motion — 36
  4.3 Undamped Free Vibration — 37
  4.4 Orthogonality — 39
  4.5 Free Vibration Due to Initial Conditions — 40
  4.6 Generalized Coordinates — 42
  4.7 Base Disturbance — 47
  4.8 Generalization to Continuous Systems — 48
  References — 49

5 **MDF SYSTEMS UNDER RANDOM LOADING** — 50
  5.1 Slender Flexible Structures — 50
  5.2 Plane Structures — 55
  5.3 Axisymmetric Structures — 59
  References — 66

6 **WIND EFFECTS ON STRUCTURES** — 68
  6.1 The Nature of Wind — 68
  6.2 Mean Wind Velocity Records — 70
  6.3 Atmospheric Turbulence — 75
  6.4 Wind Loading — 80

6.5	Response of Rectangular Structures	82
	6.5.1 Pressure Coefficients	82
	6.5.2 Pressure Spectra	85
	6.5.3 Gust Factors	89
6.6	Response of Tapered Chimneys	96
6.7	Response of Cylindrical-Type Towers	99
	6.7.1 Geometrical and Loading Characteristics	99
	6.7.2 Mean Pressure Distribution	99
	6.7.3 Fluctuating Pressure Distribution	101
6.8	Response of Suspended Roofs	115
	6.8.1 Geometrical and Loading Characteristics	115
	6.8.2 Dynamic Pressure	116
6.9	Lateral Response of Structures	119
	6.9.1 Vortex Shedding	119
	6.9.2 Galloping	122
	6.9.3 Flutter	123
	References	125
7	**EARTHQUAKE EFFECTS ON STRUCTURES**	**127**
7.1	The Nature of Earthquakes	127
	7.1.1 General Description	127
	7.1.2. Classification Systems	128
	7.1.3. Elastic Wave Model	128
7.2	Spectral Analysis	133
	7.2.1 Response Spectrum	133
	7.2.2. Inelastic Spectrum	138
	7.2.3 Design Spectrum	141
	7.2.4. Forces in MDF Systems	144
	7.2.5. Seismic Analysis of a Hyperbolic Cooling Tower	149
7.3	Time Domain Analysis	152
	7.3.1. Direct Integration of Equations of Motion	152
	7.3.2 Response of Secondary Systems	156
7.4	Systems Representation	161
	7.4.1 General	161
	7.4.2 Transfer Functions	161
	7.4.3 Source-Soil-Structure System	162
7.5	Nondeterministic Analysis	165
	7.5.1 General	165
	7.5.2 Resonance Excitation	165
	References	167
8	**CONCLUSION**	**169**
	Index	171

Preface

The response of structures to natural forces such as wind and earthquake has long been a subject of study. Only in recent times, however, have the methodologies of applied mechanics and mathematics and the prowess of digital computers enabled these problems to be approached on a fairly realistic basis, insofar as the dynamic characteristics of the structure and the random nature of the loading are concerned. Engineering practitioners are increasingly being called upon to deal with such phenomena in the course of design investigations and a grasp of the terminology and methodology is a minimum capability for a well-trained modern engineer.

This book is intended to provide the appropriate background and then to introduce the methods of analysis for structures under dynamic loading It is directed toward the practicing engineer or advanced student with a general background in structural mechanics. The material in the book has formed the basis for seminar courses attended primarily by senior level graduate students and graduate engineers.

An integrated treatment of structural response to two important types of natural loadings, wind and earthquake, seems to be warranted because of a number of common facets. In order to illustrate this, a design process chart for wind response which was prepared by a leading authority in the field, Professor Davenport*, is extended to cover earthquake response (Table 1). Although the topics are broad and perhaps superficial, the parallel is obvious.

A realistic evaluation of the structural response for such time-dependent forces draws upon several branches of mechanics and mathematics. Particularly important in this regard are the areas of structural dynamics, aerodynamics and random processes. In this treatment, the analysis of structures for dynamic loading is described within the framework of these fields.

In order to provide a sufficient background for the more advanced topics, the first two chapters are devoted to a basic description of the structural dynamics of single-degree-of-freedom (SDF) systems and to an introduction to random processes. Then the dynamic response of SDF systems to random loading is addressed. The next part of the book deals with the generalization of the mathematical description to multi-degree-of-freedom (MDF) systems. The development and utilization of realistic functions to represent wind and earthquake effects is explored, based on aerodynamic concepts. In this context general classes of engineering systems such as beams, plates and shells are considered. A special feature of this treatment is the study in some detail of four types of structures which are particularly

---

* A. G. Davenport, "Research Review: Buildings and Structures"; Wind Engineering Research Council Newsletter, Vol. 2, Dec. 1977, p. 14.

DESIGN FOR WIND RESPONSE

Climate
(wind speeds, directions, storm types)
→
Exposure
(terrain roughness, topography, local obstructions)
→
TIME HISTORY OF WIND FIELD

Structure: aerodynamics properties
(geometry: height, width, breadth)
→
TIME HISTORY OF AERODYNAMIC FORCES

Structure: mechanical properties
(stiffness, mass, damping)
→
TIME HISTORY OF STRUCTURAL RESPONSE

stress, deflection, acceleration, pressure
←
DESIGN CRITERIA

DESIGN FOR EARTHQUAKE RESPONSE

Zone
(seismicity, energy release, frequency of occurrence)
→
Local Geological and Geotechnical Features
(faults, soils, foundations)
→
TIME HISTORY OF EARTHQUAKE ACCELERATIONS

Structure: geometric properties
(configuration, framing, weight)
→
TIME HISTORY OF INERTIAL FORCES

Structure: mechanical properties
(stiffness, mass, damping)
→
TIME HISTORY OF STRUCTURAL RESPONSE

stress, deflection, acceleration, damage
←
DESIGN CRITERIA

sensitive to fluctuating loadings: tapered chimneys, hyperbolic cooling towers, suspension roofs and multi-story buildings.

The approach of the book is meant to be fairly rigorous but mainly practical. Involved mathematical proofs are avoided and the emphasis is on the use of examples to illustrate basic concepts. An attempt is made to direct the presentation to the current or anticipated state-of-the-art and to phrase the development in terminology familiar to the engineering practitioner. Ample references are cited to enable the reader to pursue any of the several topics covered in more depth. As such, the book serves as a bridge between the engineer generally trained in the techniques of structural analysis and the research specialist and should be of interest to the designer or researcher who wishes to obtain a unified view, based in structural mechanics, aerodynamics and random vibrations, of the behavior of structures under wind and earthquake loading. Examples of various classes of problems are presented; however, due to the complexity and breadth of the topics addressed, the emphasis is on the procedures rather than numerical solutions.

Phillip L. Gould
Salman H. Abu-Sitta

Acknowledgments

The authors are indebted to many colleagues and students who have served as a source of information and inspiration.
Both authors would like to recognize Professor A. G. Davenport whose pioneering contributions have made the response of structures to wind a recognized and well-defined subject and Dr C. A. Brebbia whose interest aided this combined effort. Additionally, both authors have had the opportunity to spend extended periods at the Ruhr-Universität in Bochum, West Germany and have enjoyed and benefited from contacts with Professors W. Zerna, W. Krätzig and W. Wunderlich and Drs I. Mungan, H.-J. Niemann and R. Harnach as well as many other colleagues in the Institut für Konstruktiven Ingenieurbau.
The first author would like to acknowledge the contributions of his colleague Professor T. V. Galambos and of his students, especially S. K. Sen, H. Suryoutomo and O. El-Shafee of Washington University. The support of the National Science Foundation has also enabled his research interests in this field to be sustained.
The second author would like to acknowledge the contributions of his colleagues, Professors B. J. Vickery, M. Novak, N. Isyumov, T. Jandali and D. Surry and his students, especially M. G. Hashish, J. F. Howell and I. D. Elashkar, at the University of Western Ontario. Also appreciated are the support of Professors H. Tottenham and P. B. Morice at the University of Southampton.
The manuscript was typed by Alice Bletch and Elizabeth Gould and the illustrations were produced by O. El-Shafee.

1 Structural dynamics of single-degree-of-freedom systems

1.1 GENERAL

In the vast fields of structural dynamics and mechanical vibrations, a considerable amount of the basic knowledge and the corresponding applications is based on the characteristics of simple vibrators or single-degree-of-freedom (SDF) systems. Such systems consist of three basic components, a lumped *mass*, a *spring* and a viscous *damper*, as shown in Fig. 1.1. The spring is linear and elastic in the simplest case, whereupon it develops a force proportional to the displacement of the system from the reference position. Similarly the damper produces a force proportional to the instantaneous velocity. Using d'Alembert's principle, the equation of motion for the SDF system shown in Fig. 1.1, is

$$m\ddot{x} + c\dot{x} + kx = P(t) \tag{1.1}$$

in which m, c and k are the mass, damping coefficient and spring stiffness, respectively, x is the displacement from a reference position and the applied force P is a function of time, t. Generally derivatives with respect to time are denoted by $(\cdot)$. In Fig. 1.1, the term $m\ddot{x}$ is called an inertial force.

Before taking up the questions of the characteristics and the solution of Eq. 1.1 it should be mentioned that the reduction of a complex engineering system into a relatively simple model such as the SDF system, or even to a

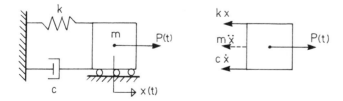

Fig. 1.1 *Single degree of freedom system*

2 STRUCTURAL DYNAMICS OF SINGLE-DEGREE-OF-FREEDOM SYSTEMS

Fig. 1.2 Elevated tank

model with more basic components, is an extremely important and highly skillful engineering discipline. Thin walled surface structures such as the elevated tank shown in Fig. 1.2 are often represented by SDF models even though there are no components physically resembling point masses or line springs. The process of modelling mathematically the dynamic response of physical systems is best explained on a case-by-case basis. Possibilities and alternatives for a few such engineering systems will be illustrated as the occasion arises.

Returning to Eq. 1.1, it is of interest to deal with two subsidiary problems:

(1) $P(t)=0, c=0$; *free* vibration
(2) $P(t)=0, c \neq 0$; *damped* free vibration

Also we may classify the original problem as

STRUCTURAL DYNAMICS OF SINGLE-DEGREE-OF-FREEDOM SYSTEMS

(3) $P(t) \neq 0$;   *forced* vibration

with the understanding that (3) is to be considered only after (1) or (2) has been solved and is thus expressed in terms of the free vibration solution. Mathematically, one could classify (1) and (2) as homogeneous solutions and (3) as a particular solution to the governing differential equation.

1.2. FREE VIBRATION

We first consider case (1), free vibration, which is described by

$$\ddot{x} + (k/m)x = 0 \tag{1.2}$$

and has the solution

$$x = C_1 \cos \bar{\omega} t + C_2 \sin \bar{\omega} t \tag{1.3a}$$

or

$$x = \sqrt{(C_1^2 + C_2^2)} \sin (\bar{\omega} t + \alpha) \tag{1.3b}$$

where

$$\bar{\omega} = \sqrt{(k/m)} \tag{1.4a}$$

$$\alpha = \tan^{-1}(C_1/C_2) \tag{1.4b}$$

and C_1 and C_2 are integration constants determined from the initial conditions as

$$C_1 = x(0) \tag{1.5a}$$

$$C_2 = \dot{x}(0)/\bar{\omega} \tag{1.5b}$$

A typical free vibration response with amplitude $A = \sqrt{(C_1^2 + C_2^2)}$ is shown in Fig. 1.3.

The parameter $\bar{\omega}$ is the *circular* natural frequency of vibration, with units of radians/second, and is related to the *actual* natural frequency $\bar{f}$, with units of cycles/second or Hz, by

$$\bar{f} = \bar{\omega}/2\pi \tag{1.6}$$

Also of interest is the fundamental *period*, the time required to complete one cycle, which is found as

$$\bar{T} = 1/\bar{f} = 2\pi/\bar{\omega} \tag{1.7}$$

4 STRUCTURAL DYNAMICS OF SINGLE-DEGREE-OF-FREEDOM SYSTEMS

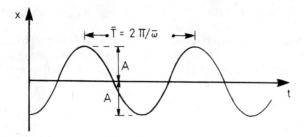

Fig. 1.3 Free vibration response of SDF system

Now considering case (2), damped free vibration, the governing equation becomes

$$\ddot{x} + (c/m)\dot{x} + (k/m)x = 0 \tag{1.8}$$

It is convenient to express the constant coefficients of Eq. 1.8 in terms of the previously defined circular natural frequency $\bar{\omega}$ as

$$c/m = 2\beta\bar{\omega} \tag{1.9a}$$

and

$$k/m = \bar{\omega}^2 \tag{1.9b}$$

where β is the so-called *coefficient of critical damping* which will be interpreted shortly.

Now with the governing equation written as

$$\ddot{x} + 2\beta\bar{\omega}\dot{x} + \bar{\omega}^2 x = 0 \tag{1.10}$$

the solution is taken as

$$x = e^{rt} \tag{1.11}$$

requiring

$$r_{1,2} = \bar{\omega}[-\beta \pm \sqrt{(\beta^2 - 1)}] \tag{1.12}$$

The most general solution of the second order equation is then

$$x = C_1 e^{r_1 t} + C_2 e^{r_2 t} \tag{1.13}$$

where again C_1 and C_2 are determined from the initial conditions.

STRUCTURAL DYNAMICS OF SINGLE-DEGREE-OF-FREEDOM SYSTEMS 5

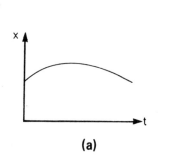
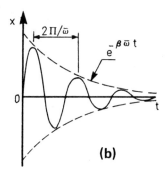

Fig. 1.4(a) *Overdamped system:* (b) *Damped free vibration response of SDF system*

The subsequent response depends on the value of the parameter β. For $\beta > 1$, $r_{1,2}$ are real and there is no oscillatory motion. The response of such a system is illustrated in Fig. 1.4(a) and may be categorized as overdamped. On the other hand for $\beta < 1$, $r_{1,2}$ are complex and Eq. 1.12 may be rewritten as

$$r_{1,2} = \bar{\omega}[-\beta \pm i\sqrt{(1-\beta^2)}] \tag{1.14}$$

and the solution as

$$x = e^{-\beta\bar{\omega}t}(C_1 \cos \bar{\omega}_d t + C_2 \sin \bar{\omega}_d t) \tag{1.15}$$

where

$$\bar{\omega}_d = \bar{\omega}\sqrt{(1-\beta^2)} \tag{1.16}$$

is the damped circular natural frequency. A typical response curve based on Eq. 1.15 is shown in Fig. 1.4(b). We see that the harmonic motion evident in Fig. 1.3 persists but with diminishing amplitude. The case $\beta = 1$, corresponding to $c = c_{cr} = 2m\bar{\omega}$, is the transition between oscillatory and non-oscillatory motion. Since $\beta = c/2m\bar{\omega}$ (Eq. 1.9a),

$$\beta = \frac{c}{c_{cr}} = \frac{c}{2\sqrt{mk}} = \frac{c}{2m\bar{\omega}} \tag{1.17}$$

which represents a fraction or percentage of critical damping. This terminology is very useful for the general classification of structural systems i.e., 0.5 to 1% = light damping; 5 to 10% = moderate damping; >15% = heavy damping, etc. Other types of damping, such as aerodynamic damping which is related to wind forces, will be considered later.

It is useful to evaluate the integration constants for Eq. 1.15 in terms of a known initial displacement and/or velocity as [1]

$$C_1 = x(0) \tag{1.18a}$$

$$C_2 = [x(0)\beta\bar{\omega} + \dot{x}(0)]/\bar{\omega}_d \tag{1.18b}$$

In many engineering systems $\beta < 0.1$ and the approximation

$$\bar{\omega}_d \simeq \bar{\omega} \tag{1.19}$$

is acceptable.

1.3 PERIODIC FORCING FUNCTION

Returning to the governing equation, Eq. 1.1, we consider the loading

$$P(t) = P_0 e^{i\omega t} = P_0(\cos \omega t + i \sin \omega t) \tag{1.20}$$

where P_0 is a constant and ω is the frequency of the forcing function. Such a loading is sometimes called a *harmonic excitation*. The general solution consists of the homogeneous solution, Eq. 1.13, plus a particular solution of the form

$$x_p = x_0 e^{i\omega t} \tag{1.21}$$

Substituting Eqs. 1.20 and 1.21 into Eq. 1.1 we find

$$x_0 = \frac{P_0}{(-\omega^2 m + i\omega c + k)} \tag{1.22}$$

Alternately, the response may be separated into static and dynamic components as

$$x_p = (P_0/k)H(\omega)e^{i\omega t} \tag{1.23}$$

where P_0/k is, in fact, the static response x_{st} and

$$H(\omega) = \frac{k}{(k + i\omega c - \omega^2 m)} \tag{1.24}$$

may be called the *complex frequency response*. It is of further interest to consider the amplitude of x_p in the form

$$\frac{|x_p|}{P_0} = \frac{|H(\omega)|}{k} \tag{1.25a}$$

STRUCTURAL DYNAMICS OF SINGLE-DEGREE-OF-FREEDOM SYSTEMS

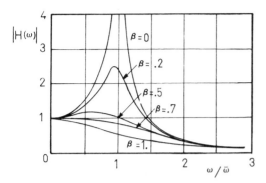

Fig. 1.5   Mechanical admittance function

where

$$|H(\omega)| = \frac{k}{[(k - \omega^2 m)^2 + \omega^2 c^2]^{1/2}} \quad (1.25b)$$

is sometimes called the *mechanical admittance function*.

One may observe that the natural frequency of the system $\bar{\omega} = \sqrt{(k/m)}$ and the damping ratio $\beta = c/(2\sqrt{(mk)})$ are contained in $|H(\omega)|$. After some algebraic manipulation we may write

$$|H(\omega)| = \frac{1}{\{[(1 - (\omega/\bar{\omega})^2]^2 + 4\beta^2(\omega/\bar{\omega})^2\}^{1/2}} \quad (1.26)$$

Equation 1.26 is extremely important in characterising the response of a system to a periodic excitation. We may refer to Fig. 1.5 where the admittance is shown as a function of the ratio of the frequencies of the excitation, ω, and the system, $\bar{\omega}$, for various damping ratios. It is seen that for $\omega/\bar{\omega} \ll 1$, $|H(\omega)| \simeq 1$ and the response

$$|x_p| = \frac{P_0}{k} \quad (1.27)$$

is essentially static or *quasistatic*. On the other hand, if $\omega/\bar{\omega} \gg 1$, $|H(\omega)| \to 0$ and the response is dominated by the mass or *inertia*. Of particular interest is $\omega/\bar{\omega} = 1$ where $|H(\omega)| = 1/2\beta$. With light damping x_p becomes very large, producing the phenomenon of *resonance*. The characteristics of the response in each regime are illustrated in Fig. 1.6.

The complete solution for the response includes the homogeneous solution, Eq. 1.15 as well. For the purpose of illustration, we consider an undamped system initially at rest ($x = \dot{x} = 0$) with the forcing function $P(t)$

8 STRUCTURAL DYNAMICS OF SINGLE-DEGREE-OF-FREEDOM SYSTEMS

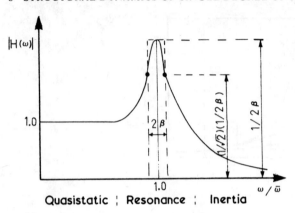

Fig. 1.6 *Characterization of system response*

Regime	Quasistatic	Resonance	Inertia
Dominant Comp.	Stiffness	Damping	Mass
$\lvert x_p \rvert / P_0$	$1/k$	$1/2\beta k = 1/\bar{\omega}c$	$1/\omega^2 m$
$\lvert \dot{x}_p \rvert / P_0$	ω/k	$1/c$	$1/\omega m$
$\lvert \ddot{x}_p \rvert / P_0$	ω^2/k	$\bar{\omega}/c$	$1/m$

$= P_0 \sin \omega t$. Using Eqs. 1.3(a) and the complex part of Eq. 1.20 and computing $H(\omega)$ from Eq. 1.26 with β taken as 0,

$$|H(\omega)| = \frac{1}{1 - (\bar{\omega}/\omega)^2} \quad (1.28)$$

Then we find

$$x_p = P_0/k \, |H(\omega)| \sin \omega t \quad (1.29a)$$

$$C_1 = 0 \quad (1.29b)$$

and

$$C_2 = -(P_0/k)(\omega/\bar{\omega})|H(\omega)| \quad (1.29c)$$

so that

$$x = (P_0/k)|H(\omega)|[-(\omega/\bar{\omega}) \sin \bar{\omega} t + \sin \omega t] \quad (1.30)$$

For the damped case the result is similar, so that we may conclude that a periodic forcing function with a natural frequency close to that of the system may produce very large responses (as compared to $x_{st} = P_0/k$) for systems which are lightly or even moderately damped. This observation is not restricted just to the case illustrated, a SDF system with a periodic forcing function. Since a more complex loading may, in fact, be represented as a Fourier series involving several forcing frequencies ω_j and a more complex system exhibits a discrete set of natural frequencies $\bar{\omega}_k$, there is the possibility of resonance and the associated amplification whenever $\omega_j \simeq \bar{\omega}_k$.

It should be noted that the previous equations are valid so long as the periodic loading continues to be applied to the system, the so-called *steady-state* phase. If the loading is terminated at a time $\bar{t}$, $x(\bar{t})$ and $\dot{x}(\bar{t})$ are computed from Eq. 1.30 and substituted into Eq. 1.18 to re-evaluate the constants which are then used in Eq. 1.15 for $t > \bar{t}$, the so-called *transient* phase.

A simple lesson in the design of systems subject to dynamic loading is to avoid resonance effects, which is ideally accomplished by choosing and arranging the structural components such that the frequencies do not match those of the forcing function. In many cases this may not be entirely possible and damping can play an important role; then the determination of a realistic value for c or β may be crucial. Also one should remember that even if some frequency matching occurs, the effect on the system may not be significant if the energy associated with the forcing function at the particular $\bar{\omega}_k$ is small.

A number of experimental procedures have been devised to calculate the damping parameters for mechanical systems[1a]. As an illustration, a curve of the form of Fig. 1.6 may be used to determine the damping contained in a system. In the figure the response of a SDF system is plotted for various excitation frequencies ω. The idea is to analyze the measured response at different ratios of $(\omega/\bar{\omega})$ to calculate β. First we use the maximum response at $\omega/\bar{\omega} = 1$ for which $|H(\omega)|$ is equal to $1/2\beta$. Next we seek the value corresponding to $\omega/\bar{\omega} = 1 + \beta$, where β is still unknown. Substituting $\omega/\bar{\omega} = 1 \pm \beta$ into Eq. 1.26 gives $H(\omega) = (1/\sqrt{2})\, 1/2\beta$ in each case, when higher powers of β are neglected. Thus, if we locate the maximum response and $(1/\sqrt{2})$ times the maximum response on both sides of resonance, the corresponding band width along the abscissa $\Delta\omega/\bar{\omega}$ is equal to 2β or

$$\beta = \tfrac{1}{2}(\Delta\omega/\bar{\omega}) \tag{1.31a}$$

Another quantity which is very useful for quantifying the damping in a mechanical test is the logarithmic decrement $\bar{\delta}$ which is the logarithm of the ratio of two successive maximum amplitudes. As a measure of β, the expression[2]

$$\bar{\delta} = 2\pi\beta/\sqrt{(1-\beta^2)} \simeq 2\pi\beta \tag{1.31b}$$

is widely used.

1.4 RESPONSE TO IMPULSE AND CONVOLUTION

Consider a impulsive loading of the form

$$P(t) = I\delta(t) \tag{1.32a}$$

in which I is the magnitude of the impulse and $\delta(t)$ is the Dirac δ-function defined such that $\delta(t) = 0$ for all $t \ne 0$ but

$$\int_{0-}^{0+} \delta(t) dt = 1 \tag{1.32b}$$

If the system is initially at rest, Newton's second law indicates that an initial velocity

$$\dot{x}(0) = I/m \tag{1.33}$$

is produced.

Substituting $x(0) = 0$ and Eq. 1.33 into Eqs. 1.18 and then into 1.15 and using Eqs. 1.9(b) and 1.16 gives the response as

$$x = \frac{I}{k}\frac{\bar{\omega}}{\sqrt{(1-\beta^2)}} e^{-\beta\bar{\omega}t} \sin \bar{\omega}_d t \tag{1.34}$$

It is convenient to write Eq. 1.34 in the form

$$x = \frac{I}{k} h(t) \tag{1.35a}$$

where

$$h(t) = \frac{\bar{\omega}}{\sqrt{(1-\beta^2)}} e^{-\beta\bar{\omega}t} \sin \bar{\omega}_d t \tag{1.35b}$$

is called *the unit impulse function*.

Now we consider a force $P(t)$ which is divided into a sequence of impulses, Fig. 1.7. The magnitude of the impulse at a time $t = \tau$ is equal to $P(\tau)d\tau$ and the associated response at time t is

$$x(t) = P(\tau)/k \ \ h(t-\tau)d\tau \tag{1.36}$$

since h is a function of the difference between the time of observation t and the time of application τ. The total response to all such impulses is given by

STRUCTURAL DYNAMICS OF SINGLE-DEGREE-OF-FREEDOM SYSTEMS 11

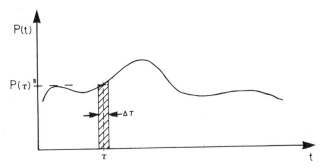

Fig. 1.7 Time dependent loading

$$x(t) = \int_{-\infty}^{t} P(\tau)/k \; h(t-\tau) d\tau \qquad (1.37)$$

since the system is affected by only those impulses applied before time t. Equation 1.37 may also be written as[3]

$$x(t) = \int_{0}^{\infty} P(t-\tau)/k \; h(\tau) d\tau \qquad (1.38)$$

Equations 1.37 and 1.38 are forms of the convolution integral or Duhamel's integral.

For the case of the periodic forcing function given by Eq. 1.20

$$P(t-\tau) = P_0 e^{i\omega t} e^{-i\omega \tau} \qquad (1.39)$$

and

$$x_p(t) = (P_0/k) e^{i\omega t} \int_{0}^{\infty} h(\tau) e^{-i\omega \tau} d\tau \qquad (1.40)$$

is a particular solution. Upon comparison with Eq. 1.23, we identify

$$H(\omega) = \int_{0}^{\infty} h(\tau) e^{-i\omega \tau} d\tau \qquad (1.41)$$

indicating that the complex frequency response is the Fourier transform of

12 STRUCTURAL DYNAMICS OF SINGLE-DEGREE-OF-FREEDOM SYSTEMS

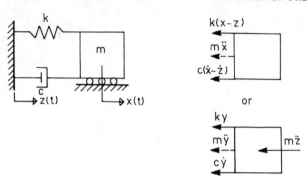

Fig. 1.8 SDF system with base motion

the unit impulse function. This relationship is quite important in the cases of random loading which are treated subsequently.

1.5 BASE DISTURBANCE MODEL

We again consider the SDF model shown in Fig. 1.1 with an additional coordinate $z(t)=$ the motion of the base as defined in Fig. 1.8. While the inertial term in Eq. 1.1 remains proportional to the *absolute displacement* x, the damping and spring forces are assumed to be proportional to the *relative displacement* $y=x-z$. Therefore the equation of motion in the absence of other forces is

$$m\ddot{x}+c\dot{y}+ky=0 \tag{1.42}$$

Subtracting $m\ddot{z}$ from both sides and clearing, we have

$$\ddot{y}+2\beta\bar{\omega}\dot{y}+\bar{\omega}^2 y = -\ddot{z} \tag{1.43}$$

as the governing equation. This equation is widely used in the analysis of SDF systems due to earthquake motion. It is evident that the response of a SDF system with a base acceleration equal to $\ddot{z}$ is identical to the forced vibration of a fixed base system with a forcing function of $-m\ddot{z}$. The time dependent response is then written from Eq. 1.38 as

$$y(t) = \int_0^\infty -\ddot{z}(t-\tau)/\bar{\omega}^2 h(\tau)\mathrm{d}\tau \tag{1.44}$$

STRUCTURAL DYNAMICS OF SINGLE-DEGREE-OF-FREEDOM SYSTEMS 13

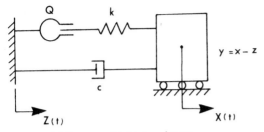

Fig. 1.9 SDF system with friction element

1.6 ELASTO-PLASTIC MODEL

For some structures it is unfeasible to provide resistance to extreme dynamic forces totally by elastic action and the energy absorbtion capability or *ductility* of the system must be utilized to supplement the elastic capacity.

A simple extension of the SDF model shown in Fig. 1.8 to accommodate elasto-plastic action is shown in Fig. 1.9. In series with the linear spring is a Coulomb friction element Q which is defined such that

$$Q = ky \text{ if } ky < Q_Y \quad (1.45)$$

$$Q = Q_Y \text{ if } ky \geqslant Q_Y \quad (1.46)$$

where

$$Q_Y = k y_Y = \text{Yield strength of spring} \quad (1.47)$$

The resulting force-displacement diagram after several excursions into the inelastic regime is shown in Fig. 1.10. The maximum displacement is related to the yield displacement y_Y by

$$\mu = y_{\max}/y_Y = \text{ductility} \quad (1.48)$$

Equation 1.43 is generalized to

$$\ddot{y} + 2\beta\bar{\omega}\dot{y} + q = -\ddot{z} \quad (1.49)$$

in which

$$q = Q/m \quad (1.50)$$

It is convenient to introduce the change of variables

$$v = \dot{y} \quad (1.51)$$

STRUCTURAL DYNAMICS OF SINGLE-DEGREE-OF-FREEDOM SYSTEMS

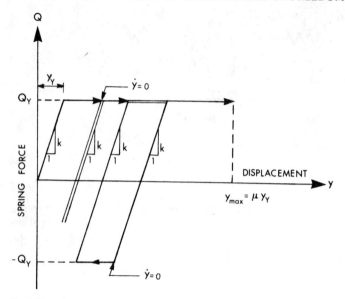

Fig. 1.10   Elasto-plastic force displacement diagram

whereupon Eq. 1.49 can be rewritten as

$$\dot{v} = -(\ddot{z} + 2\beta\bar{\omega}v + q) \tag{1.52}$$

which may be integrated numerically as an initial value problem[1b] for v and again for y. Of course, the value of q will have to be assumed and later checked at each stage.

After the response $y(t)$ has been calculated for the duration of the ground motion $z(t)$, the maximum response is substituted into Eq. 1.48 to determine the required ductility ratio μ. It is implied that the structural system can be designed and constructed to develop this ductility and that the accompanying non-recoverable deformation can be accepted.

REFERENCES

1. Clough, R. W. and Penzien, J., *Dynamics of Structures*, McGraw-Hill, New York, 1975, p. 46: (a) pp. 69–77; (b) pp. 118–128
2. Jacobsen, L. S. and Ayre, R. S., *Engineering Vibrations*, McGraw-Hill, New York, 1958, pp. 200–203
3. Robson, J. D., *An Introduction to Random Vibrations*, Edinburgh University Press, U.K., 1963, p. 56

2 Random processes

2.1 GENERAL

We consider a continuing occurrence, perhaps the deflection of a structure under wind load or the straining of a seawall under wave action and, for the moment, assume that the action is described by a single variable x which is a function of time. The process is sampled over a reasonable interval as shown in Fig. 2.1. It is assumed that the interval is long enough so that the essential characteristics of the signal $x(t)$ within the sample period, such as the mean and extreme values, would be much the same for a different sampling period. For such processes, which are quite common in nature, it is often meaningful to characterize $x(t)$ as a *random variable* and to use the techniques of statistical analysis to describe the process. The precise assumptions and requirements associated with *random processes* will be given later.

It is of interest to distinguish between random processes which are sampled *continuously*, such as that shown in Fig. 2.1, and those for which $x(t)$ is measured at *discrete* time intervals.

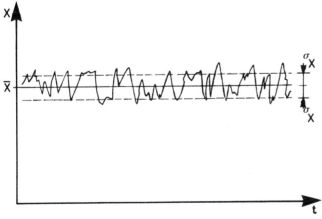

Fig. 2.1   Random variable

2.2 STATISTICAL PARAMETERS

The *mean* value of x over a time period $\bar{T}$ is

$$\bar{x} = \frac{1}{\bar{T}} \int_0^{\bar{T}} x(t) dt \qquad (2.1a)$$

for a continuous variable and

$$\bar{x} = \frac{1}{\bar{T}} \sum_{t=0}^{\bar{T}} x(t) \Delta t \qquad (2.1b)$$

for a discrete variable. It is common in the literature to call the mean value of the random variable $E[x(t)]$, the *expectation*, and also to use the notation $\langle x(t) \rangle$ for the time average of a function. Thus $\bar{x} = E[x(t)] = \langle x(t) \rangle$. If the argument in Eq. 2.1(a) is $x^2(t)$, then $\langle x^2(t) \rangle$ is known as the *mean square* value of $x(t)$.

After establishing the mean, we are interested in describing fluctuations about this value. In order to eliminate algebraic signs it is convenient to use the square of the deviation from the mean $[x(t) - \bar{x}(t)]^2$ as a measure of the fluctuations and to define the mean of this quantity as the *variance* σ_x^2:

$$\sigma_x^2 = \langle [x(t) - \bar{x}(t)]^2 \rangle \qquad (2.2)$$

σ_x is known as the *standard deviation* which indicates the average spread of x about $\bar{x}$. Of course, the variance is equal to the mean square when $\bar{x} = 0$. Equation 2.2 may be written as[1]

$$\sigma_x^2 = \langle x^2(t) \rangle - \langle x(t) \rangle^2 \qquad (2.3a)$$

One is often interested in comparing the magnitude of the spread to the mean, which is given by the dimensionless variance

$$\bar{\sigma}_x^2 = \sigma_x^2(x) / \langle x(t) \rangle^2 \qquad (2.3b)$$

$\bar{\sigma}_x$ is called the *coefficient of variation*.

2.3 PROBABILITY

We are interested in those values of a random variable $x(t)$ which are less than or equal to a particular value of x. For example, we may measure the deflections of a mast over a representative period $\bar{T} = m\Delta t$ at each Δt and obtain, respectively $x_0, x_2, \ldots, x_m$. For any given value of x, we may count the number of $x_i \leq x$, say j. Then the ratio $j/(m+1)$ gives the probability that $x_i \leq x$. That is to say

$$P(x_i(t) \leq x) = P(x) \qquad (2.4)$$

RANDOM PROCESSES 17

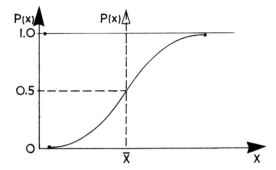

Fig. 2.2 Probability distribution function

$P(x)$ is known as the *probability distribution function* and a typical curve is plotted in Fig. 2.2. We observe that $0 < P(x) < 1$ and is symmetrical about the mean $x = \bar{x}$. It is straightforward to write the probability of a given $x(t)$ lying between two bounds, say x_1 and x_2, as

$$P(x_1 \leq x(t) \leq x_2) = P(x_2) - P(x_1) \tag{2.5}$$

Moving to the differential level,

$$\Delta P(x) = P(x \leq x(t) \leq x + \Delta x) = P(x + \Delta x) - P(x) \tag{2.6}$$

which may be regarded as proportional to Δx, i.e.,

$$\Delta P(x) = p(x)\Delta x \tag{2.7}$$

In the limit as $\Delta x \to 0$

$$p(x) = \frac{dP(x)}{dx} \tag{2.8}$$

which is called the *probability density function* as shown in Fig. 2.3. Conversely, if $p(x)$ is known

$$P(x) = \int_{-\infty}^{x} p(x)dx \tag{2.9}$$

Also it is clear that

$$\int_{-\infty}^{\infty} p(x)dx = 1 \tag{2.10}$$

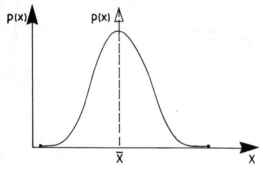

Fig. 2.3 Probability density function

The shape of the function $p(x)$ qualitatively describes the nature of the distribution. A narrow, peaked curve indicates close clustering of the variable about the mean while a broader, rounded curve represents more scatter.

The previously defined statistical parameters may be conveniently expressed in terms of $p(x)$:

Mean:

$$\bar{x} = \int_{-\infty}^{\infty} x p(x) \mathrm{d}x \qquad (2.11)$$

which is termed the *first moment* of the area of the probability density curve about $p(x)$.

Variance:

$$\sigma_x^2 = \int_{-\infty}^{\infty} [x - \bar{x}]^2 p(x) \mathrm{d}x \qquad (2.12)$$

which is called the *second moment*.

In structural dynamics, it is often convenient to treat the fluctuations only since the mean response is largely static. This is easily synthesized from the foregoing by setting $\bar{x} = 0$, which corresponds to the dashed ordinate scales in Figs. 2.2 and 2.3. The variance for this case

$$\sigma_x^2 = \langle \dot{x}^2(t) \rangle = \int_{-\infty}^{\infty} x^2 p(x) \mathrm{d}x \qquad (2.13)$$

is called the *mean square* of $x(t)$ while the standard deviation σ_x is *the root-mean-square* (RMS).

Although the definition of the probability density function implies that a large number of observations of a *given* process have been recorded, there is a similarity in the pattern of occurrence of natural events and a possibility to transfer one set of observations, presumably well-documented, to a different process which may be less documented. In engineering analysis, it is common to assume *a priori* the form of $p(x)$ which is most often Gaussian[2]:

$$p(x) = 1/(\sigma_x \sqrt{2\pi}) e^{-(x-\bar{x})^2/(2\sigma_x^2)} \qquad (2.14)$$

or

$$p(x) = 1/(\sigma_x \sqrt{2\pi}) e^{-x^2/(2\sigma_x^2)} \qquad (2.15)$$

for the zero mean case. The shape of $p(x)$ is controlled by σ_x, a small value producing a narrow, peaked curve and a large value a broad, rounded curve.

2.4 MULTIPLE RANDOM VARIABLES

In the monitoring of an event it is common to obtain a number of records, e.g., the deflections at several stations on the structure. A set of such records is called an *ensemble*. Assuming that we still have a single independent variable, and that each record is of a different form, e.g., $x_{(1)}(t)$, $x_{(2)}(t) \ldots x_{(m)}(t)$; where $x_{(1)}(t) \neq x_2(t) \neq x_{(m)}(t)$, the probability density function is theoretically a function of all of the variables $p(x_{(1)}, x_{(2)} \ldots x_{(i)} \ldots x_{(n)})$. Such a multivariate form complicates the analysis substantially and it is common in engineering to consider only one or two components of the ensemble[2a].

For two random variables, say $x_{(1)} = x$, $x_{(2)} = y$, we require some additional statistical parameters. The *joint probability density function* refers to the three dimensional analogues of Figs. 2.2 and 2.3 such that

$$p(x, y) = \frac{\partial^2 P(x, y)}{\partial x \partial y} \qquad (2.16)$$

while

$$\int_{-\infty}^{\infty} \int_{-\infty}^{\infty} p(x, y) \, dx \, dy = 1 \qquad (2.17)$$

or, conversely,

$$P(x, y) = \int_{-\infty}^{x} \int_{-\infty}^{y} p(x, y) dx dy \qquad (2.18)$$

In addition to the variances of x and y which are written in analogous form to Eqs. 2.2, 2.3(a) and 2.12

$$\sigma_x^2 = \langle [x(t) - \bar{x}(t)]^2 \rangle$$
$$= \langle x^2(t) \rangle - \langle x(t) \rangle^2$$
$$= \int_{-\infty}^{\infty} \int_{-\infty}^{\infty} [x - \bar{x}]^2 p(x, y) dx dy \qquad (2.19)$$

and

$$\sigma_y^2 = \langle [y(t) - \bar{y}(t)]^2 \rangle$$
$$= \langle y^2(t) \rangle - \langle y(t) \rangle^2$$
$$= \int_{-\infty}^{\infty} \int_{-\infty}^{\infty} [y - \bar{y}]^2 p(x, y) dx dy \qquad (2.20)$$

We also have the *covariance*:

$$\sigma_{xy} = \langle [x(t) - \bar{x}(t)][y(t) - \bar{y}(t)] \rangle$$
$$= \langle x(t) y(t) \rangle - \langle x(t) \rangle \langle y(t) \rangle$$
$$= \int_{-\infty}^{\infty} \int_{-\infty}^{\infty} [x - \bar{x}][y - \bar{y}] p(x, y) dx dy \qquad (2.21)$$

Additionally we have the *correlation coefficient*

$$\rho_{xy} = \sigma_{xy}/(\sigma_x \sigma_y) \qquad (2.22)$$

If the random variables are statistically independent,

$$\langle xy \rangle = \langle x \rangle \langle y \rangle$$

RANDOM PROCESSES 21

and both σ_{xy} and ρ_{xy} are zero. The assumption of statistical independence is frequently invoked to simplify engineering analyses. The generalization to n variables is straight forward[2b].

Considering the ensemble of records

$$\{x(t)\} = \{x_{(1)}(t) x_{(2)}(t) \ldots x_{(i)}(t) \ldots x_{(n)}(t)\}$$

at various instants in time, say $t_1, t_2, \ldots, t_j \ldots t_m$ and computing the statistical parameters of each record at each instant, we may establish the corresponding time variation for each $x_{(i)}$, $(i = 1, n)$. If, however, these parameters remain constant with time, the process is said to be *stationary*.

If we calculate the statistical parameters for a single record $x_{(i)}(t)$ at a particular time, say t_j, and also those for the ensemble $\{x(t)\}$ at the same t_j, we may establish the corresponding record deviation. If, however, the statistical parameters for the single record are equal to those for the ensemble, the process is said to be *ergodic*.

Simply, a stationary process does not depend on the time origin while an ergodic process does not depend on which record is selected. Our definitions imply that an ergodic process must always be stationary, while a stationary process may or may not be ergodic[2c]. When a process is considered to be both stationary and ergodic, the statistical properties computed for a representative interval of a single record $x(t)$ can be used to represent the ensemble $\{x(t)\}$. This assumption will generally be used throughout the book.

2.5 AUTOCORRELATION AND POWER SPECTRA

We now consider a random process described by the record $x(t)$ and examine the relationship between $x(t)$ and $x(t + \tau)$ where τ is a time delay. It is obvious that as τ becomes smaller, $x(t + \tau) \to x(t)$ but it is not necessarily obvious what happens as τ increases. It is apparent that the relationship between $x(t)$ and $x(t + \tau)$ for various values of τ should give a measure of statistical dependence or correlation of the two variables, and the *autocorrelation function*

$$R_x(\tau) = \langle x(t) x(t + \tau) \rangle \quad (2.23a)$$

is conveniently defined as the covariance of the variable for the zero mean case.

It is sometimes more convenient to write R_x is the form of a limit as

$$R_x(\tau) = \lim_{T \to \infty} \frac{1}{2T} \int_{-T}^{T} x(t) x(t + \tau) dt \quad (2.23b)$$

22 RANDOM PROCESSES

We are able to offer some observations about $R_x(\tau)$. We have previously shown that $R_x(0) = \langle x^2(t) \rangle$, the mean square value of $x(t)$. If $\tau > 0$ but still relatively small, most products can be expected to be positive and $0 < R_x(\tau) < R_x(0)$. As τ increases, two random functions become increasingly unrelated until $R_x(\infty) = 0$. The rate at which $R_x(\tau) \to 0$ indicates the pattern of the fluctuations of $x(t)$ with a waveform exhibiting rapid oscillations showing a faster decline in $R_x(\tau)$ than a slowly varying curve. The sensitivity of $R_x(\tau)$ to the pattern of oscillations indicates a frequency dependence which will be developed later.

We now consider the fluctuating component of the waveform ($\bar{x} = 0$) as represented by the Fourier integral

$$x(t) = \frac{1}{2\pi} \int_{-\infty}^{\infty} c(\omega) e^{i\omega t} d\omega \tag{2.24}$$

with

$$c(\omega) = \int_{-\infty}^{\infty} x(t) e^{-i\omega t} dt \tag{2.25}$$

Next we consider the function $x^2(t)$ and evaluate

$$\int_{-\infty}^{\infty} x^2(t) dt = \int_{-\infty}^{\infty} x(t) \frac{1}{2\pi} \int_{-\infty}^{\infty} c(\omega) e^{i\omega t} d\omega$$

$$= \frac{1}{2\pi} \int_{-\infty}^{\infty} c(\omega) c^*(i\omega) d\omega$$

$$= \frac{1}{\pi} \int_{0}^{\infty} |c(\omega)|^2 d\omega \tag{2.26}$$

where $c^*(\omega)$ is the complex conjugate of $c(\omega)$. Equation 2.26 is similar to the mean square of $x^2(t)$ except that the infinite limits are awkward. To overcome this, a new function x_T is defined such that

$$x_T(t) = x(t), \quad -\frac{\bar{T}}{2} < t < \frac{\bar{T}}{2} \tag{2.27}$$

$$= 0, \text{ all other } t$$

where $\bar{T}$ = the fundamental period as defined by Eq. 1.7. Now the mean square of $x_T^2(t)$ is written from Eqs. 2.1(a) and 2.26 as

$$\langle x_T^2(t) \rangle = \frac{1}{\bar{T}} \int_{-\bar{T}/2}^{\bar{T}/2} x_T^2(t) dt$$

$$= \frac{1}{\bar{T}} \int_{-\infty}^{\infty} x_T^2(t) dt$$

$$= \frac{1}{\bar{T}\pi} \int_0^{\infty} |c_T(\omega)|^2 d\omega \qquad (2.28)$$

If we now let the period $\bar{T} \to \infty$ we obtain $\langle x^2(t) \rangle$:

$$\sigma_x^2 = \langle x^2(t) \rangle = \frac{1}{\pi} \int_0^{\infty} \lim_{\bar{T} \to \infty} \left[\frac{1}{\bar{T}} |c_T(\omega)|^2 \right] d\omega$$

$$= \int_0^{\infty} S_x(\omega) d\omega \qquad (2.29)$$

where

$$S_x(\omega) = \lim_{\bar{T} \to \infty} \left[\frac{1}{\pi \bar{T}} |c_T(\omega)|^2 \right] \qquad (2.30)$$

is called the *power spectral density* of the function $x(t)$.

We may interpret $S_x(\omega)$ in a general way. Considering a band of width $\Delta\omega$ in Eq. 2.29, the corresponding portion of $\langle x^2(t) \rangle$ is $S_x(\omega)\Delta\omega$, which leads to the more formal definition of $S_x(\omega)\Delta\omega$ as the mean square value of the signal passed by a narrow-band filter of band width $\Delta\omega^{1a}$. The phase independence of the modulus $c_T(\omega)$ means that $S_x(\omega)$ as determined for a particular function $x(t)$ may be applicable to a range of similar functions. A simple example is the case of $S_x(\omega) =$ constant for all ω which is termed *white noise*. In practice, the use of such a signal must be accompanied by a cut-off frequency so that Eq. 2.29 will remain finite. Finally we note that for a Gaussian process with zero mean, as given by Eq. 2.15, $p(x)$ is completely defined by the power spectral density since the variance $\sigma_x^2 = \langle x^2(t) \rangle$, as given by Eq. 2.29.

We have previously alluded to the dependence of the autocorrelation function $R_x(\tau)$ on the frequency content of the record which implies a connection to the power spectral density $S_x(\omega)$. Before developing this relationship, it is of interest to note that these two functions provide alternative means of describing the dynamic response of systems. For now, it is sufficient to state that an analysis based on $R_x(\tau)$ is said to be in the *time domain* while an analysis based on $S_x(\omega)$ is in the *frequency domain*.

The autocorrelation function and the power spectral density are a Fourier transform pair[2d]:

$$R_x(\tau) = \frac{1}{2} \int_{-\infty}^{\infty} S_x(\omega) e^{i\omega\tau} d\omega \qquad (2.31)$$

and

$$S_x(\omega) = \frac{1}{\pi} \int_{-\infty}^{\infty} R_x(\tau) e^{-i\omega\tau} d\tau \qquad (2.32)$$

One should be cautious in using these equations, as they specifically pertain to $S_x(\omega)$ as defined by Eqs. 2.29 and 2.30. Some references extend the definition over the negative frequency range as well[2e] and others[1b] use the natural frequency f instead of the circular natural frequency ω. It is helpful in the latter case to remember that $S_x(f) = 2\pi S_x(\omega)$ and $df = d\omega/2\pi$.

The effect of these and similar variations is to merely change the constants which multiply the integrals.

Using the fact that $S_x(\omega)$ is an even function, the range of integration may be confined to positive time and frequencies by using[1c]

$$R_x(\tau) = \int_0^\infty S_x(\omega) \cos \omega\tau \, d\omega \qquad (2.33)$$

and

$$S_x(\omega) = \frac{2}{\pi} \int_0^\infty R_x(\tau) \cos \omega\tau \, d\tau \qquad (2.34)$$

The relationship between the autocorrelation function and the power spectral density enables one to conveniently move between the time and frequency domains. For example, the input function may be measured in terms of time while the system characteristics may be evaluated in terms of frequency dependent functions.

It is also of interest to note that the autocorrelation function of the 'white noise' signal mentioned previously is the Dirac-δ function as defined following Eq. 1.31(b)[1d].

2.6 COMBINED RANDOM PROCESSES

We sometimes deal with random processes which are made up of components, each of which is a randomly varying quantity. We will take the elementary case of two superposed stationary variables

$$z(t) = x(t) + y(t) \tag{2.35}$$

The autocorrelation function for the combined variable is

$$R_z(\tau) = \langle z(t)z(t+\tau) \rangle \tag{2.36}$$

and is given by

$$R_z(\tau) = R_x(\tau) + R_y(\tau) + R_{yx}(\tau) + R_{xy}(\tau) \tag{2.37}$$

where R_x and R_y pertain to the variables x and y alone and are defined by Eq. 2.23 with the appropriate argument, and R_{yx} and R_{xy} are *cross-correlation functions* given by

$$R_{yx}(\tau) = \langle x(t)y(t+\tau) \rangle \tag{2.38}$$

$$R_{xy}(\tau) = \langle y(t)x(t+\tau) \rangle \tag{2.39}$$

The cross-correlation functions are related by

$$R_{xy}(\tau) = R_{yx}(-\tau) \tag{2.40}$$

and do not necessarily have maxima at $\tau = 0$. Of course if x and y are unrelated, $R_{yx} = R_{xy} = 0$.

Corresponding to the cross-correlation functions are *cross-spectral densities*[1e] or *cross-spectra*

$$S_{xy}(\omega) = \frac{1}{\pi} \int_{-\infty}^{\infty} R_{xy}(\tau) e^{-i\omega\tau} d\tau \tag{2.41}$$

and

$$S_{yx}(\omega) = \frac{1}{\pi} \int_{-\infty}^{\infty} R_{yx}(\tau) e^{-i\omega\tau} d\tau \qquad (2.42)$$

The cross-spectral densities are complex so that

$$S_{xy}(\omega) = C_{xy}(\omega) + iQ_{xy}(\omega) \qquad (2.43)$$

and

$$S_{yx}(\omega) = C_{yx}(\omega) + iQ_{yx}(\omega) \qquad (2.44)$$

where C_{xy} and C_{yx} are called *co-spectra and* Q_{xy} and Q_{yx} are termed *quadratures.* The quadrature represents the cross-spectrum if x or y is shifted in phase $90°$ at frequency ω. Also S_{xy} and S_{yx} are related by

$$S_{xy}(\omega) = S_{yx}{}^*(\omega) \qquad (2.45)$$

and

$$S_{xy}(\omega) = S_{yx}(-\omega) \qquad (2.46)$$

Finally we define the *coherence function* or *cross-correlation coefficient*

$$\gamma_{xy}^2(\omega) = \frac{|S_{yx}(\omega)|^2}{S_x(\omega) S_y(\omega)} < 1.0 \qquad (2.47)$$

where

$$|S_{xy}(\omega)|^2 = C_{xy}^2(\omega) + Q_{xy}^2(\omega) \qquad (2.48)$$

which establishes the frequency dependence of the spectra.

If x and y are physical locations on a structure, γ_{xy} would tend to describe the effect of spatial separation as well. This is useful when a response at two different locations, say x_1 and x_2, at the same time t is evaluated. Considering Eq. 2.29, we may compute

$$\sigma_{x_1 x_2} = \langle x_1 x_2 \rangle = \int_0^\infty S_{x_1 x_2}(\omega) d\omega \qquad (2.49)$$

whereby

$$\gamma^2_{x_1 x_2} = \frac{|S_{x_1 x_2}|}{S_{x_1} S_{x_2}} \qquad (2.50)$$

REFERENCES

1. Robson, J. D., *An Introduction to Random Vibrations*, Edinburgh University Press, U.K., 1963, p.19: (a) p. 35; (b) pp. 40–41; (c) p. 41; (d) p. 42; (e) pp. 44–47.
2. Clough, R. W. and Penzien, J., *Dynamics of Structures*, McGraw-Hill, New York, 1975, pp. 401–403: (a) p. 437; (b) pp. 430–431; (c) pp. 443–444; (d) pp. 451–457; (e) p. 452.

3 SDF systems under random loading

3.1 SIMPLE OSCILLATOR

We consider the SDF system shown in Fig. 1.1 subject to a loading $P(t)$. The solution was generated in the general form of a convolution integral as Eq. 1.37. Now, we evaluate the autocorrelation function of the response by substituting Eq. 1.37 into Eq. 2.23(b)

$$R_x(\tau) = \frac{1}{k^2} \lim_{T \to \infty} \frac{1}{2T} \int_{-T}^{T} \int_{0}^{\infty} h(\tau_1) P(t-\tau_1) d\tau_1 \int_{0}^{\infty} h(\tau_2) \cdot$$

$$\cdot P(t+\tau-\tau_2) d\tau_2 dt$$

$$= \frac{1}{k^2} \int_{0}^{\infty} \int_{0}^{\infty} h(\tau_1) h(\tau_2) \lim_{T \to \infty} \frac{1}{2T} \int_{-T}^{T} P(t-\tau_1) \cdot$$

$$\cdot P(t+\tau-\tau_2) dt d\tau_1 d\tau_2 \qquad (3.1)$$

If we compare the last integral in Eq. 3.1 with Eq. 2.23(b), we may express

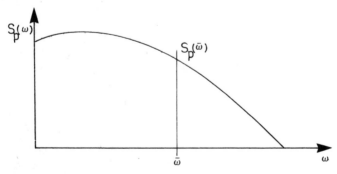

Fig. 3.1 Wind pressure spectrum

the argument of the first term, $P(t-\tau_1)$, as $P(\bar{t})$ with $\bar{t}=t-\tau_1$, whereupon the argument of the second term, $P(t+\tau-\tau_2)$, becomes $P(\bar{t}+\tau_1+\tau-\tau_2)=P(\bar{t}+\tau+\tau_1-\tau_2)$. Since $dt=d\bar{t}$, the integral becomes

$$R_P(\tau+\tau_1-\tau_2)=\lim_{T\to\infty}\frac{1}{2T}\int_{-T}^{T}P(\bar{t})P(\bar{t}+\tau+\tau_1-\tau_2)d\bar{t} \qquad (3.2)$$

which is the autocorrelation of the forcing function. Now Eq. 3.1 may be written as

$$R_x(\tau)=[G_P^x]R_P(t+\tau_1-\tau_2) \qquad (3.3)$$

where the autocorrelation of the response is related to the autocorrelation of the excitation by the *transfer function*

$$[G_P^x]=\frac{1}{k^2}\int_0^\infty\int_0^\infty h(\tau_1)h(\tau_2)d\tau_1 d\tau_2 \qquad (3.4)$$

The concept of a transfer function is helpful in describing the behaviour of complex dynamic systems since $[G_P^x]$ may be visualized as a 'black box' which converts the input P to the output x and thus embodies the characteristics of the entire system, however complex they may be. One can imagine that the isolation of an explicit closed form transfer function for a complex system may be beyond reach but the notion is nevertheless still useful in the separation of such systems into manageable components for analysis. In the case at hand, it is apparent from Eq. 1.35b that $[G_P^x]$ is a linear filter which is a function of the system parameters only ($\bar{\omega}$, β and k) and is thus applicable for any form of applied loading.

The system may be described in the frequency domain as well. Starting with the basic transformation over the interval $-\infty<\tau<\infty$ given by Eq. 2.32,

$$S_x(\omega)=\frac{1}{\pi}\int_{-\infty}^{\infty}R_x(\tau)e^{-i\omega\tau}d\tau \qquad (3.5)$$

Then, substituting Eq. 3.3 and 3.4 into 3.5[1],

$$S_x(\omega) = \frac{1}{\pi k^2} \int_{-\infty}^{\infty} \left[\int_0^{\infty} \int_0^{\infty} h(\tau_1)d\tau_1 h(\tau_2)d\tau_2 \right.$$

$$\left. \cdot R_P(\tau + \tau_1 - \tau_2)d\tau_1 d\tau_2 \right] e^{-i\omega\tau} d\tau \qquad (3.6)$$

Next the order and limits of integration are altered to produce

$$S_x(\omega) = \frac{1}{\pi k^2} \lim_{T \to \infty} \left[\int_0^T h(\tau_1)d\tau_1 \int_0^T h(\tau_2)d\tau_2 \int_{-T}^T R_P(\tau + \tau_1 - \tau_2)e^{-i\omega\tau}d\tau \right]$$

which may be rearranged as

$$S_x = \frac{1}{\pi k^2} \lim_{T \to \infty} \left[\int_0^T h(\tau_1)e^{i\omega\tau_1}d\tau_1 \cdot \int_0^T h(\tau_2)e^{-i\omega\tau_2}d\tau_2 \right.$$

$$\left. \cdot \int_{-T}^T R_P(\tau + \tau - \tau_2)e^{-i\omega(\tau + \tau_1 - \tau_2)}d\tau \right] \qquad (3.7)$$

Now, considering Eqs. 1.25(b), 1.26, 1.41 and 2.32

$$S_x(\omega) = \frac{1}{\pi k^2} H^*(\omega)H(\omega) \cdot \pi S_P(\omega) = \frac{1}{k^2} |H(\omega)|^2 S_P(\omega) \qquad (3.8)$$

in which the complex frequency response

$$H(\omega) = \frac{1}{[1 + 2i\beta(\omega/\bar{\omega}) - (\omega/\bar{\omega})^2]} \qquad (3.9a)$$

and $|H(\omega)|$ = the mechanical admittance function previously evaluated in Eq. 1.26. Computationally

$$|H(\omega)|^2 = [H(\omega)H^*(\omega)]$$

$$= \frac{1}{\{[1 - (\omega/\bar{\omega})^2]^2 + 4\beta^2(\omega/\bar{\omega})^2\}} \qquad (3.9b)$$

is more convenient.

In many references, the constant $1/k^2$ is incorporated into $H(\omega)^2$ but, for

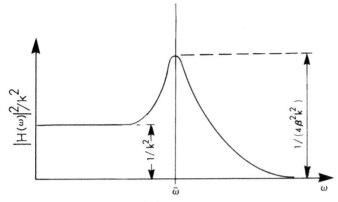

Fig. 3.2 Transfer function for SDF system

our purposes, it is convenient to keep this term separate and to define the additional function $\alpha = H(\omega)/k$ when the notation becomes cumbersome.

Equation 3.8 expresses the relationship between response and excitation in the frequency domain. Noting the similarity to the solution of a SDF system under *harmonic* excitation given in Chapter 1, we observe that the transfer function $(1/k^2)|H(\omega)|^2$ is the square of the ratio of the maximum displacement to excitation force for the harmonic excitation of a SDF system. The square power is attributable to the definition of $S(\omega)$ as the variance or mean square per frequency, which is evident from the integrant of Eq. 2.29.

Thus, in a given frequency interval $\Delta\omega$, the variance of response produced by a variance of force is a harmonic phenomena.

3.2 RESPONSE OF SDF SYSTEM TO WIND

As an illustration, we consider a billboard represented as a SDF system with viscous damping. A general discussion of this type of system is presented by Davenport[3]. For our purposes we may consider the idealized wind pressure spectrum shown in Fig. 3.1. In this case, the 'spectrum' is simply a plot of $S_p(\omega)$ against ω. Figure 3.2 is a plot of the transfer function $(1/k^2)|H(\omega)|^2$. This curve is the square of Fig. 1.6 divided by k^2. Note the characteristic amplification in the vicinity of $\bar{\omega}$. The (displacement) response spectrum is obtained as the product of Fig. 3.1 and 3.2 and is shown in Fig. 3.3.

Writing the variance of the response from Eqs. 2.29 and 3.8,

$$\sigma_x^2 = \int_0^\infty \frac{|H(\omega)|^2}{k^2} S_p(\omega) d\omega \qquad (3.10)$$

32 SDF SYSTEMS UNDER RANDOM LOADING

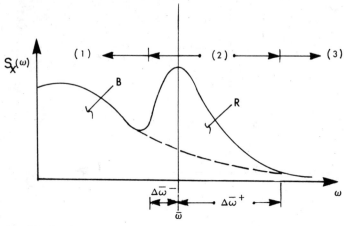

Fig. 3.3 *Response spectrum for a billboard*

Equation 3.10 can be clarified by breaking the response spectrum into three regions:

(1) $0 < \omega < \bar{\omega} - \Delta\bar{\omega}^-$;
(2) $\bar{\omega} - \Delta\bar{\omega}^- < \omega < \bar{\omega} + \Delta\bar{\omega}^+$ and
(3) $\bar{\omega} + \Delta\bar{\omega}^+ < \omega$.

The lines of delineation are shown qualitatively in Fig. 3.3 where the bounds for $\Delta\bar{\omega}^\pm$ are exaggerated for clarity.

Obviously region (3) contributes little and may be neglected while region (1) gives

$$\sigma^2_{x(1)} = \int_0^{\bar{\omega}-\Delta\bar{\omega}^-} |H(\omega)|^2/k^2 \cdot S_P(\omega) d\omega \qquad (3.11)$$

which is dependent on the stiffness of the system only since $|H(\omega)|^2$ is essentially constant and equal to 1. ($H(\omega)$ is carried in the following expressions to facilitate generalization to cases which are dimensional.) The response in this region is called *quasistatic*.

The most interesting region is (2) where

$$\sigma^2_{x(2)} = \int_{\bar{\omega}-\Delta\bar{\omega}^-}^{\bar{\omega}+\Delta\bar{\omega}^-} |H(\omega)|^2/k^2 \cdot S_P(\omega) d\omega \qquad (3.12)$$

since $H(\omega)$ is strongly dependent on $\bar{\omega}$ and β.

It is common to approximate the resonance component as $\pi/2$ times the area of the dashed rectangle in Fig. 1.6. Applying this to the product of Figs. 3.1 and 3.2 gives

$$\sigma_R^2 = (\pi/2) \cdot 2\beta\bar{\omega} \cdot 1/(4\beta^2 k^2) \cdot S_P(\bar{\omega})$$

$$= (\pi/4\beta) \cdot \bar{\omega}/k^2 \cdot S_P(\bar{\omega}) \qquad (3.13a)$$

which was used by Abu-Sitta and Davenport[4].
In terms of the displacement spectra, Eq. 3.13(a) may be written as

$$\sigma_R^2 = \pi\beta\bar{\omega}S_x(\bar{\omega}) \qquad (3.13b)$$

in view of Eq. 3.8, with $|H(\omega)|^2/k^2 = 1/(4\beta^2 k^2)$ as shown in Fig. 3.2. Thus we have

$$\sigma_x^2 = \sigma_B^2 + \sigma_R^2 \qquad (3.14)$$

in which

$$\sigma_B^2 = |H(\omega)|^2/k^2 \int_0^{\bar{\omega}} S_P(\omega)d\omega \qquad (3.15a)$$

with

$$|H(\omega)|^2 = 1 \qquad (3.15b)$$

from Eq. 3.11 and σ_R^2 given by Eq. 3.13.

The separation of the variance into quasistatic and resonant parts is valuable in many engineering applications. Another useful approach, the gust factor method, is introduced in Sections 5.3.

3.3 MULTIPLE LOADS

Considering the two random forces $P(t)$ and $Q(t)$, the computation of the response $x(t)$ is a lengthy algebraic procedure. Starting with Eq. 3.1 and adding the correlated contributions of P and Q, Robson[2a] shows that the spectrum is

$$S_x(\omega) = \alpha_P^* \alpha_P S_P(\omega) + \alpha_P^* \alpha_Q S_{PQ}(\omega)$$

$$+ \alpha_Q^* \alpha_P S_{QP}(\omega) + \alpha_Q^* \alpha_Q S_Q(\omega) \qquad (3.16)$$

in which

$$\alpha_j = H_j(\omega)/k_j \quad (j = P, Q) \tag{3.17}$$

and S_{PQ} and S_{QP} are cross-spectra given by Eqs. 2.41 and 2.42, respectively, with $x = P$ and $y = Q$. If P and Q are uncorrelated

$$S_x(\omega) = |\alpha_P|^2 S_P(\omega) + |\alpha_Q|^2 S_Q(\omega) \tag{3.18}$$

and if the correlation is proportional such that $Q(t) = \lambda P(t)^{2b}$

$$R_{PQ}(\tau) = R_{QP}(\tau) = \lambda R_P(\tau) \tag{3.19}$$

and

$$R_Q(\tau) = \lambda^2 R_P(\tau) \tag{3.20}$$

so that Eq. 3.16 becomes

$$S_x(\omega) = |\alpha_P + \lambda \alpha_Q|^2 S_P(\omega) \tag{3.21}$$

The generalization to n correlated forces P_i is easily written as

$$S_x(\omega) = \sum_{i=1}^{n} \sum_{j=1}^{n} \alpha_{xP_i} \alpha_{xP_j} S_{P_i P_j}(\omega) \tag{3.22}$$

3.4 EXTENSIONS

There are a number of further possibilities with respect to cross-correlations. Two displacements, say x and y, due to a single force may be described in a relative sense only if their cross spectral densities or cross-correlation functions are known[2c]. Additionally, the excitation and response, or cause and effect, may be cross-correlated. Any generalities which are introduced into the model for purposes of improving the realism are accompanied by a need for sufficient data to evaluate the accompanying cross-correlations so that a theoretically elegant stochastic model may be difficult to quantify.

REFERENCES

1. Clough, R. W. and Penzien, J., *Dynamics of Structures*, McGraw-Hill, New York, 1975, p. 492.
2. Robson, J. D., *An Introduction to Random Vibrations*, Edinburgh University Press, 1963, p. 58: (a) pp. 74–77; (b) p. 78; (c) p. 87.

3. Davenport, A. G., "The Application of Statistical Concepts to the Wind Loading of Structures", *Proc. Inst, civ. Engrs.*, London, Vol. 19, 1961, pp. 449–472.
4. Abu-Sitta, S. H. and Davenport, A. G., "Earthquake Design of Cooling Towers", *J. Struct. Div. ASCE*, Vol. 96, No. ST9, Sept. 1970, pp. 1889–1902.

4 Structural dynamics of MDF systems

4.1 GENERAL

Multiple degrees of freedom may be grouped into two distinct types of systems: (1) Those in which the mass and stiffness properties are *distributed* with respect to the spatial coordinates of the system; and (2) Those for which the mass and stiffness properties are *discretized* at a reasonable number of coordinates or node points. We shall have occasion to consider both types but, from the standpoint of analysis, the discrete parameter models are far more general and adaptable to automatic computation and will be pursued in this chapter. In those cases where the displacements are required to be in the form of continuous functions, we will indicate the appropriate transformations.

Our format treatment will be restricted to problems involving a single spatial coordinate and appropriate generalizations will be indicated as they arise.

4.2 EQUATIONS OF MOTION

A discrete model subject to time dependent forces as shown in Fig. 4.1 is described by the system of equations

$$[M]\{\ddot{X}\}+[C]\{\dot{X}\}+[K]\{X\}=\{P\} \tag{4.1}$$

in which $[M]$, $[C]$, $[K]$ and $\{P\}$ are the mass, damping and stiffness matrices and the load vector, respectively, and $\{X\}$ is the displacement

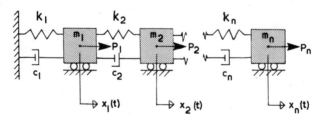

Fig. 4.1 Discrete MDF system

vector or *geometric coordinate* vector. Although Eq. 4.1 is symbolically equivalent to Eq. 1.1, several comments are in order regarding the generalization to n coordinate points or *nodes*.

First, the mass matrix $[M]$ will be diagonal if the masses are simply lumped at the nodes. This has some computational advantages and is adequate for many cases. For systems which are geometrically complex such as curved beams or shells, more precision can be obtained using a variationally *consistent* mass matrix which is non-diagonal. This is discussed in depth in books dealing with finite element analysis.

The damping matrix $[C]$ is likewise diagonal and is conveniently chosen to be a linear combination of $[M]$ and $[K]$ for reasons which will be obvious later.

The stiffness matrix $[K]$ may be assembled using the direct stiffness method. The component member stiffness matrices are available for a wide range of structural elements.

The load vector $\{P\}$ consists of one element per coordinate. Again physical lumping is often used to represent distributed loading but consistent load vectors may be advantageous for more complex systems.

4.3 UNDAMPED FREE VIBRATION

The free vibration solution will be derived neglecting $[C]$ by assuming harmonic motion with

$$\{X\} = \{\varphi\}e^{i\omega t} \tag{4.2}$$

giving the linear eigenvalue problem

$$[K]\{\varphi\} = \omega^2[M]\{\varphi\} \tag{4.3}$$

which may be solved routinely for diagonal or even non-diagonal mass matrices[1].

The solution to Eq. 4.3 consists of a set of eigenvalues or natural frequencies

$$\{\Omega\} = \{\bar{\omega}_1 \bar{\omega}_2 \ldots \bar{\omega}_j \ldots \bar{\omega}_n\} \tag{4.4}$$

and an associated mode shape or eigenvector

$$\{\varphi^j\} = \{\varphi_1^j \varphi_2^j \ldots \varphi_n^j\} \tag{4.5}$$

for each $\bar{\omega}_j$. The mode shapes may be collected as

$$[\Phi] = [\{\varphi^1\}\{\varphi^2\} \ldots \{\varphi^j\} \ldots \{\varphi^n\}] \tag{4.6}$$

38 STRUCTURAL DYNAMICS OF MDF SYSTEMS

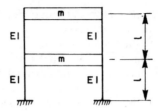

Fig. 4.2 2DF frame structure

Since the governing equation is homogeneous, each eigenvector is solved only within a constant.

Then the complete solution is written from Eq. 4.2 as

$$\{X\} = \sum_{j=1}^{n} b_j\{\varphi^j\} \cos \bar{\omega}_j t + d_j\{\varphi^j\} \sin \bar{\omega}_j t \qquad (4.7)$$

where the constants b_j and d_j are elements of

$$\{B\} = \{b_1 b_2 \ldots b_j \ldots b_n\} \qquad (4.8a)$$

and

$$\{D\} = \{d_1 d_2 \ldots d_j \ldots d_n\} \qquad (4.8b)$$

The constants are determined from the initial conditions which may be specified either at $t=0$ or $t=\bar{t}$, which corresponds to the time of removal of a forcing function and the onset of transient vibration.

As an example we introduce a two story, fixed base frame with equal masses m, story heights l and flexural rigidities EI, as shown in Fig. 4.2. For this structure $k_1 = k_2 = k = 2 \times 12EI/l^3 = 24EI/l^3$ and

$$[M] = \begin{bmatrix} m & 0 \\ 0 & m \end{bmatrix} \qquad (4.9)$$

$$[K] = \begin{bmatrix} 2k & -k \\ -k & k \end{bmatrix} \qquad (4.10)$$

leading to the eigenvalues

$$\{\Omega\} = \{\sqrt{0.382} \ \sqrt{2.62}\} \sqrt{k/m} \qquad (4.11)$$

and the eigenvectors

$$[\Phi] = \begin{bmatrix} 1.000 & 1.000 \\ 1.618 & -0.618 \end{bmatrix} \qquad (4.12)$$

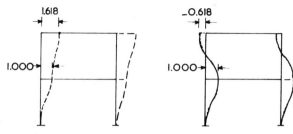

Fig. 4.3 *Mode shapes for a 2DF frame*

where φ_1^j is taken as one. The mode shapes are shown in Fig. 4.3. The complete solution is written as

$$\begin{Bmatrix} x_1 \\ x_2 \end{Bmatrix} = b_1 \begin{Bmatrix} 1.000 \\ 1.618 \end{Bmatrix} \cos \sqrt{(0.382\ k/m)}\, t +$$

$$+ b_2 \begin{Bmatrix} 1.000 \\ -0.618 \end{Bmatrix} \cos \sqrt{(2.62\ k/m)}\, t$$

$$+ d_1 \begin{Bmatrix} 1.000 \\ 1.618 \end{Bmatrix} \sin \sqrt{(0.382\ k/m)}\, t$$

$$+ d_2 \begin{Bmatrix} 1.000 \\ -0.618 \end{Bmatrix} \sin \sqrt{(2.62\ k/m)}\, t \qquad (4.13)$$

4.4 ORTHOGONALITY

When the natural frequencies and mode shapes are computed using the undamped equations of motion, the modes of vibration are orthogonal with respect to both the stiffness and mass matrices. This is expressed by[2]

$$\{\varphi^i\}^T [K]\{\varphi^j\} = 0 \qquad (i \ne j) \qquad (4.14)$$

and

$$\{\varphi^i\}^T [M]\{\varphi^j\} = 0 \qquad (i \ne j) \qquad (4.15)$$

For our 2DF system

$$\lfloor 1.000\ -0.618 \rfloor \begin{bmatrix} 2 & -1 \\ -1 & 1 \end{bmatrix} \begin{Bmatrix} 1.000 \\ 1.618 \end{Bmatrix} k = 0 \qquad (4.16a)$$

$$\lfloor 1.000 - 0.618 \rfloor \begin{bmatrix} 1 & 0 \\ 0 & 1 \end{bmatrix} \begin{Bmatrix} 1.000 \\ 1.618 \end{Bmatrix} m = 0 \qquad (4.16b)$$

4.5 FREE VIBRATION DUE TO INITIAL CONDITIONS

If the system is subject to a set of initial displacements

$$\{X(0)\} = \{\bar{X}\} \qquad (4.17)$$

and/or initial velocities

$$\{\dot{X}(0)\} = \{\dot{\bar{X}}\} \qquad (4.18)$$

the constants of integration are chosen to satisfy the equations

$$\{\bar{X}\} = \sum_{j=1}^{n} \bar{b}_j \{\varphi^j\} \qquad (4.19)$$

and

$$\{\dot{\bar{X}}\} = \sum_{j=1}^{n} \bar{d}_j \bar{\omega}_j \{\varphi^j\} \qquad (4.20)$$

which are obtained from Eq. 4.7 and d/dt of Eq. 4.7 evaluated at $t=0$.

To determine the individual constants for any mode i, both sides of each equation are premultiplied by $[M]$ and $\{\varphi^i\}^T$ to give, in view of the orthogonality relations,

$$\{\varphi^i\}^T[M]\{\bar{X}\} = \sum_{j=1}^{n} \bar{b}_j \{\varphi^i\}^T[M]\{\varphi^j\}$$

$$= \bar{b}_i \{\varphi^i\}^T[M]\{\varphi^i\} \qquad (4.21)$$

and

$$\{\varphi^i\}^T[M]\{\dot{\bar{X}}\} = \sum_{j=1}^{n} \bar{d}_j \bar{\omega}_j \{\varphi^i\}^T[M]\{\varphi^j\}$$

$$= \bar{d}_i \bar{\omega}_i \{\varphi^i\}^T[M]\{\varphi^i\} \qquad (4.22)$$

Thus, the constants are given by

$$\bar{b}_i = \frac{\{\varphi^i\}^T[M]\{\bar{X}\}}{\{\varphi^i\}^T[M]\{\varphi^i\}} \qquad (4.23)$$

STRUCTURAL DYNAMICS OF MDF SYSTEMS 41

and

$$\bar{d}_i = \frac{\{\varphi^i\}^T[M]\{\dot{\bar{X}}\}}{\bar{\omega}_i\{\varphi^i\}^T[M]\{\varphi^i\}} \qquad (4.24)$$

which may be assembled into $\{\bar{B}\}$ and $\{\bar{D}\}$, respectively (Eq. 4.8). Finally, Eq. 4.7 becomes

$$\{X\} = \sum_{j=1}^{n} \bar{b}_j\{\varphi^j\} \cos \bar{\omega}_j t + \bar{d}_j\{\varphi^j\} \sin \bar{\omega}_j t \qquad (4.25)$$

It is of interest to note that Eq. 4.25, which describes the free vibration of an nDF system under a set of initial conditions, consists of three distinct parts:

$\bar{b}_j, \bar{d}_j =$ Participation factors for mode j

$\{\varphi^j\} =$ Mode shape for mode j

$\cos \bar{\omega}_j t, \sin \bar{\omega}_j t =$ Generalized displacements for mode j.

Again referring to the 2DF example, we consider an initial displacement

$$\{X(0)\} = \{\bar{X}\} = \begin{Bmatrix} 1 \\ 1 \end{Bmatrix} \qquad (4.26)$$

and zero initial velocity. Substituting Eq. 4.26 into Eq. 4.23 gives

$$\bar{b}_1 = \frac{\lfloor 1.000 \quad 1.618 \rfloor \begin{bmatrix} 1 & 0 \\ 0 & 1 \end{bmatrix} \begin{Bmatrix} 1 \\ 1 \end{Bmatrix} m}{\lfloor 1.000 \quad 1.618 \rfloor \begin{bmatrix} 1 & 0 \\ 0 & 1 \end{bmatrix} \begin{Bmatrix} 1.000 \\ 1.618 \end{Bmatrix} m} = 0.72 \qquad (4.27a)$$

and

$$\bar{b}_2 = \frac{\lfloor 1.000 \ -0.618 \rfloor \begin{bmatrix} 1 & 0 \\ 0 & 1 \end{bmatrix} \begin{Bmatrix} 1 \\ 1 \end{Bmatrix} m}{\lfloor 1.000 \ -0.618 \rfloor \begin{bmatrix} 1 & 0 \\ 0 & 1 \end{bmatrix} \begin{Bmatrix} 1.000 \\ -0.618 \end{Bmatrix} m} = 0.28 \qquad (4.27b)$$

Then Eq. 4.25 becomes

$$\begin{Bmatrix} x_1 \\ x_2 \end{Bmatrix} = 0.72 \begin{Bmatrix} 1.000 \\ 1.618 \end{Bmatrix} \cos \sqrt{(0.382 \ k/m)} t + 0.28 \begin{Bmatrix} 1.000 \\ -0.618 \end{Bmatrix} \cos \sqrt{(2.62 \ k/m)} t$$

$$(4.28)$$

42 STRUCTURAL DYNAMICS OF MDF SYSTEMS

As a second example consider

$$\{\bar{X}\} = \begin{Bmatrix} 1.000 \\ 1.618 \end{Bmatrix} = \{\varphi^1\}$$

Then $\bar{b}_1 = 1.0$ and $\bar{b}_2 = 0$ indicating that if the system is displaced initially into one of its natural modes, it will vibrate exclusively in that mode.

4.6 GENERALIZED COORDINATES

From the previous discussion and specifically from Eq. 4.7, it is apparent that the displacement of the system at a given time may be represented by the superposition of the corresponding displacements of each normal mode at that instant.

The contribution of the individual modes may be represented as the product of the *mode shape* and the *amplitude*, i.e.

$$\{X\} = [\Phi]\{Q\} \qquad (4.29)$$

in which the time dependent amplitudes

$$\{Q\} = \{q_1 q_2 \ldots q_j \ldots q_n\} \qquad (4.30)$$

are called *generalized coordinates* or *normal coordinates*.

It is of value to rewrite the equations of motion in generalized coordinates. Substituting Eq. 4.29 into Eq. 4.1 and premultiplying by an arbitrary mode shape $\{\varphi^i\}^T$ we have

$$\{\varphi^i\}^T[M][\Phi]\{\ddot{Q}\} + \{\varphi^i\}^T[C]\{\Phi\}\{\dot{Q}\}$$

$$+ \{\varphi^i\}^T[K][\Phi]\{Q\} = \{\varphi^i\}^T\{P\} \qquad (4.31)$$

We note the orthogonality conditions, Eq. 4.14 and 4.15, which give

$$\{\varphi^i\}^T[M][\Phi]\{\ddot{Q}\} = \lfloor 0\ 0 \ldots \bar{m}_i \ldots 0 \rfloor\{\ddot{Q}\} = \bar{m}_i \ddot{q}_i \qquad (4.32a)$$

in which

$$\bar{m}_i = \{\varphi^i\}^T[M]\{\varphi^i\} \qquad (4.32b)$$

$$= \text{Generalized mass for mode } i$$

and

$$\{\varphi^i\}^T[K][\Phi]\{Q\} = [0\ 0\ ...\ \bar{k}_i\ ...\ 0]\{Q\}$$

$$= \bar{k}_i q_i \quad (4.33a)$$

in which

$$\bar{k}_i = \{\varphi^i\}^T[K]\{\varphi^i\} \quad (4.33b)$$

$$= \text{Generalized stiffness for mode } i$$

As previously mentioned, the damping is taken as a linear combination of $[M]$ and $[K]$

$$[C] = \alpha_1[M] + \alpha_2[K] \quad (4.34)$$

so that it is orthogonal as well and

$$\{\varphi^i\}^T[C][\Phi]\{\dot{Q}\} = [0\ 0\ ...\ \bar{c}_1\ ...\ 0]\{\dot{Q}\}$$

$$= \bar{c}_i \dot{q}_i \quad (4.35a)$$

in which

$$\bar{c}_i = \{\varphi^i\}^T[C]\{\varphi^i\}$$

$$= \text{Generalized damping for mode } i \quad (4.35b)$$

The evaluation of α_1 and α_2 will be discussed later.
Finally

$$\{\varphi^i\}^T\{P\} = \bar{p}_i$$

$$= \text{Generalized load for mode } i \quad (4.36)$$

Noting Eqs. 4.32–4.36, Eq. 4.31 reduces to a system of n *uncoupled* equations

$$\bar{m}_i \ddot{q}_i + \bar{c}_i \dot{q}_i + \bar{k}_i q_i = \bar{p}_i \quad (i = 1, ..., n) \quad (4.37)$$

which are identical to Eq. 1.1.
It is remarkable that the MDF system can be reduced to a set of uncoupled SDF systems, and the transformation from geometric to generalized coordinates is surely among the most elegant in mechanics.

44 STRUCTURAL DYNAMICS OF MDF SYSTEMS

Carrying the formulation one step further, we divide through by $\bar{m}_i$ to obtain

$$\ddot{q}_i + 2\beta_i \bar{\omega}_i \dot{q}_i + \bar{\omega}_i^2 q_i = \bar{p}_i/\bar{m}_i \quad (i=1,\ldots,n) \quad (4.38)$$

which is essentially identical to Eq. 1.10.
The undamped circular natural frequency for mode i is

$$\bar{\omega}_i = \sqrt{(\bar{k}_i/\bar{m}_i)} \quad (4.39)$$

The $\bar{\omega}_i$ are of course identical to those in Eq. 4.4.
Also we have introduced a modal damping coefficient

$$\beta_i = \frac{\bar{c}_i}{(2\bar{m}_i \bar{\omega}_i)}$$

$$= \frac{\bar{c}_i}{2\sqrt{(\bar{m}_i \bar{k}_i)}} \quad (4.40)$$

Noting Eq. 4.34,

$$\beta_i = \frac{1}{2}(\alpha_1/\bar{\omega}_i + \alpha_2 \bar{\omega}_i) \quad (4.41)$$

so that α_1 and α_2 can be chosen to give a specified β_i at two different frequencies. Also this form admits the possibility of using a different β_i for each mode. This type of damping, known as Rayleigh damping, is convenient and some useful generalizations are possible[2a].

The solution to Eq. 4.38 for an arbitrary loading is written from the Duhamel integral in the form of Eq. 1.37 with $h(t)$ evaluated from Eq. 1.35(b), damping neglected and $t>0$:

$$q_i(t) = \frac{1}{\bar{m}_i \bar{\omega}_i} \int_0^t \bar{p}_i(\tau) e^{-\beta_i \bar{\omega}_i (t-\tau)} \sin \bar{\omega}_i(t-\tau) d\tau$$

$$(i=1, n) \quad (4.42)$$

If the initial displacement and/or velocity are not zero, a free vibration solution in the form of Eq. 4.25 must be added. It is, of course, convenient to write that solution in generalized coordinates as well. Since the initial conditions are given in terms of the geometric coordinates, an appropriate transformation is required.

Starting with Eq. 4.29 and premultiplying by $\{\varphi^i\}^T[M]$, we find

$$\{\varphi^i\}^T[M]\{X\} = \{\varphi^i\}^T[M][\Phi]\{Q\}$$

$$= \bar{m}_i q_i \qquad (4.43)$$

so that

$$q_i = 1/\bar{m}_i \{\varphi^i\}^T[M]\{X\} \qquad (4.44)$$

and

$$\dot{q}_i = 1/\bar{m}_i \{\varphi^i\}^T[M]\{\dot{X}\} \qquad (4.45)$$

For a set of initial conditions $\{\bar{X}\}$ and $\{\dot{\bar{X}}\}$ specified in geometric coordinates, the corresponding values in generalized coordinates are

$$\{\bar{Q}\} = [\bar{M}]^{-1}[\Phi]^T[M]\{\bar{X}\} \qquad (4.46)$$

and

$$\{\dot{\bar{Q}}\} = [\bar{M}]^{-1}[\Phi]^T[M]\{\dot{\bar{X}}\} \qquad (4.47)$$

in which

$$[\bar{M}] = \lceil \bar{m}_i \rfloor \qquad (4.48)$$

Then we may write the free vibration response for each generalized coordinate q_i using Eqs. 1.15 and 1.18 as

$$q_i(t) = e^{-\beta_i \bar{\omega}_i t} [(\dot{\bar{q}}_i + \bar{q}_i \beta_i \bar{\omega}_i)/\bar{\omega}_i \sin \bar{\omega}_i t$$

$$+ \bar{q}_i \cos \bar{\omega}_i t] \qquad (i = 1, n) \qquad (4.49)$$

again neglecting damping.

Finally the response in geometric coordinates for any time, $\bar{t}$, is computed by back-substitution of Eqs. 4.6 and the sum of Eqs. 4.42 and 4.49, evaluated at $\bar{t}$, into Eq. 4.29.

We note that the elastic forces in the system are given, in general, by

$$\{F(t)\} = [K]\{X(t)\} = [K][\Phi]\{Q(t)\} \qquad (4.50)$$

46 STRUCTURAL DYNAMICS OF MDF SYSTEMS

To continue with the 2DF example shown in Fig. 4.2, we first compute

$$\bar{m}_1 = \lfloor 1.000 \ \ 1.618 \rfloor \begin{bmatrix} 1 & 0 \\ 0 & 1 \end{bmatrix} \begin{Bmatrix} 1.000 \\ 1.618 \end{Bmatrix} m$$

$$= 3.62m \tag{4.51a}$$

$$\bar{m}_2 = \lfloor 1.000 \ \ -0.618 \rfloor \begin{bmatrix} 1 & 0 \\ 0 & 1 \end{bmatrix} \begin{Bmatrix} 1.000 \\ -0.616 \end{Bmatrix} m$$

$$= 1.38m \tag{4.51b}$$

$$\bar{k}_1 = \lfloor 1.000 \ \ 1.618 \rfloor \begin{bmatrix} 2 & -1 \\ -1 & 1 \end{bmatrix} \begin{Bmatrix} 1.000 \\ 1.618 \end{Bmatrix} k$$

$$= 1.38k \tag{4.52a}$$

$$\bar{k}_2 = \lfloor 1.000 \ \ -0.618 \rfloor \begin{bmatrix} 2 & -1 \\ -1 & 1 \end{bmatrix} \begin{Bmatrix} 1.000 \\ -0.618 \end{Bmatrix} k$$

$$= 3.62k \tag{4.52b}$$

and

$$[\bar{M}] = \begin{bmatrix} 3.62 & 0 \\ 0 & 1.38 \end{bmatrix} m \tag{4.53}$$

We verify the natural frequencies given by Eq. 4.11 by using Eq. 4.39

$$\bar{\omega}_1 = \sqrt{(1.38k/3.62m)} = \sqrt{(0.382k/m)} \tag{4.54a}$$

$$\bar{\omega}_2 = \sqrt{(3.62k/1.38m)} = \sqrt{(2.62k/m)} \tag{4.54b}$$

This comparison is valuable in checking the arithmetic computations. We also transform the initial displacement vector given by Eq. 4.26 into

$$\{\bar{Q}\} = \begin{bmatrix} 1/3.62 & 0 \\ 0 & 1/1.38 \end{bmatrix} \begin{bmatrix} 1.000 & 1.618 \\ 1.000 & -0.618 \end{bmatrix} \begin{bmatrix} 1 & 0 \\ 0 & 1 \end{bmatrix} \begin{Bmatrix} 1 \\ 1 \end{Bmatrix} = \begin{Bmatrix} 0.72 \\ 0.28 \end{Bmatrix}$$

$$\tag{4.55}$$

Then, for $\beta_i = 0$

$$q_1 = 0.72 \cos \sqrt{(0.382k/m)}t \qquad (4.56a)$$

$$q_2 = 0.28 \cos \sqrt{(2.62k/m)}t \qquad (4.56b)$$

whereupon

$$\{X\} = \begin{Bmatrix} x_1 \\ x_2 \end{Bmatrix} = \begin{bmatrix} 1.000 & 1.000 \\ 1.618 & -0.618 \end{bmatrix} \begin{Bmatrix} 0.72 \cos \sqrt{(0.382k/m)}t \\ 0.28 \cos \sqrt{(2.62k/m)}t \end{Bmatrix} \qquad (4.57)$$

which is identical to Eq. 4.28.

4.7 BASE DISTURBANCE

By analogy with Eq. 1.43, the effective loading at mass point j is

$$p_j(t) = -m_j \ddot{z}(t) \qquad (4.58)$$

where $\ddot{z}(t)$ = the base acceleration. The loading vector for the nDF system is therefore

$$\{P\} = -[M]\{I\}\ddot{z}(t) \qquad (4.59)$$

where $\{I\}$ is a unit vector.
The generalized loading for mode i is written from Eq. 4.36 as

$$\bar{p}_i = \{\varphi^i\}^T[M]\{I\}\ddot{z}(t)$$

$$= -\hat{p}_i \ddot{z}(t) \qquad (4.60)$$

where $\hat{p}_i$ is the participation factor for mode i.
Then the right hand side of Eq. 4.38 becomes

$$\bar{p}_i/\bar{m}_i = -(\hat{p}_i/\bar{m}_i)\ddot{z}(t) \qquad (4.61)$$

and the modal response is evaluated by Eq. 4.42.
Finally, the total response is computed from Eq. 4.29 with $\{X\}$ replaced by the relative displacement vector $\{Y\}$:

$$\{Y\} = [\Phi]\{Q\} \qquad (4.62)$$

4.8 GENERALIZATION TO CONTINUOUS SYSTEMS

In the consideration of the response of structures, it is sometimes advantageous to have the mode shapes or eigenvectors as well as the loading represented as continuous functions over the surface of the structure. If the system is relatively simple, one may obtain the normal modes by solving the governing equation of the system or through an approximate method such as Rayleigh's method for generalized SDF systems[2b] or the more formal Rayleigh–Ritz method for complex systems[2c]. Readily available are solutions for prismatic beams with various boundary conditions[3]. On the other hand, if the structure is analyzed as a discrete system, a continuous $\{\varphi^i\}$ may be obtained by using a finite Fourier series or a polynomial approximation. For a one dimensional system described by a coordinate ξ, $0<\xi<L$, the generalized mass, Eq. 4.32(b) becomes[2d]

$$\bar{m}_i(t) = \int_0^L [\varphi^i(\xi)]^2 m(\xi) d\xi \qquad (4.63)$$

and the generalized loading, Eq. 4.36, becomes

$$\bar{p}_i(t) = \int_0^L \varphi^i(\xi) p(\xi, t) d\xi \qquad (4.64)$$

The generalized stiffness is somewhat more complicated and is easiest written from Eq. 4.39 as

$$\bar{k}_i(t) = \bar{\omega}_i^2 \bar{m}_i \qquad (4.65)$$

assuming that $\bar{\omega}_i$ has been evaluated.

As an example we consider the 2DF system shown in Fig. 4.2 and represent $\varphi^1 = \{1.000 \ \ 1.618\}$ as $\varphi(\xi)$, $(0<\xi<2l)$. We choose a sine series

$$\varphi(\xi) = \sum_{k=1}^{3} d_k \sin(k\pi\xi/4l) \qquad (4.66)$$

which identically satisfies $\varphi(0) = 0$. The Fourier coefficients are determined from the conditions

$$\varphi'(0) = 0 = (\pi/4l)d_1 + (2\pi/4l)d_2 + (3\pi/4l)d_3$$

$$\varphi(l) = 1.0 = d_1 \sin(\pi/4) + d_2 \sin(\pi/2) + d_3 \sin(3\pi/4)$$

$$\varphi(2l) = 1.618 = d_1 \sin(\pi/2) + d_2 \sin\pi + d_3 \sin(3\pi/2)$$

from which

$$d_1 = 0.483, \quad d_2 = 1.461, \quad d_3 = -1.135$$

and

$$\varphi(\xi) = 0.483 \sin(\pi\xi/4l) + 1.461 \sin(\pi\xi/2l)$$
$$- 1.135 \sin(3\pi\xi/4l) \qquad (4.67)$$

which can be used in Eq. 4.63 with $L = 2l$. A polynomial function would also be satisfactory.

REFERENCES

1. Gupta, K. K., "Eigenvalue Solution by a Combined Sturm Sequence and Inverse Iteration Technique", *Int. J. Numerical Meth. Engng*, Vol. 7, 1973, pp. 17–42.
2. Clough, R. W. and Penzien, J., *Dynamics of Structures*, McGraw-Hill, New York, 1975, pp. 185–186: (a) pp. 194–199; (b) pp. 29–34; (c) pp. 239–243; (d) p. 332
3. Jacobsen, L. S. and Ayre, R. S., *Engineering Vibrations*, McGraw-Hill, New York, 1958, pp. 482–496.

5 MDF systems under random loading

5.1 SLENDER FLEXIBLE STRUCTURES

As opposed to systems which can be described by a single point mass, natural frequency and mode shape, slender systems with distributed mass such as beams and cables possess an infinite number of degrees of freedom and corresponding modes of vibration. Also the spatial variations of the loading along the structure may be important.

We consider the loading to be a fluctuating pressure $p(\xi, t)$ with zero mean. If the mean is not zero, the mean pressure is applied as a static loading and the resulting stresses and deflections are added to the dynamic values. We assume that a normal mode analysis has been performed and the mode shapes are expressed as continuous functions $\varphi^i(\xi, t)$ for each mode as discussed in Section 4.8. The autocorrelation of the generalized loading, Eq. 4.64, is written from Eq. 2.23(b) as

$$R_{p_i}(\tau) = \lim_{T \to \infty} \frac{1}{2T} \int_{-T}^{T} \bar{p}_i(t) \bar{p}_i(t+\tau) dt$$

$$= \int_0^L \int_0^L \varphi^i(\xi_1) \varphi^i(\xi_2) d\xi_1 d\xi_2 \lim_{T \to \infty} \frac{1}{2T} \int_{-T}^{T} p(\xi, t) p(\xi, t+\tau) dt$$

$$= \int_0^L \int_0^L R_p(\xi_1, \xi_2, \tau) \varphi^i(\xi_1) \varphi^i(\xi_2) d\xi_1 d\xi_2 \tag{5.1}$$

in which L = the length of the member.

Equation 5.1 is dependent on the autocorrelation of the applied loading $R_p(\xi_1, \xi_2, \tau)$, the *space-time correlation function*, which reflects the pressure correlation between points ξ_1 and ξ_2. If the points are close together, the correlation should be strong while if the points are far apart, the correlation would be weak; thus R_p is expected to be a rapidly decaying function.

MDF SYSTEMS UNDER RANDOM LOADING 51

The mean square or variance of the generalized loading is

$$\sigma_{p_i}^2 = \int_0^L \int_0^L \langle p_i(\xi_1, t) p_i(\xi_2, t) \varphi^i(\xi_1) \varphi^i(\xi_2) d\xi_1 d\xi_2 \rangle \tag{5.2}$$

where the integrand represents the covariance of $p(\xi, t)$ at points ξ_1 and ξ_2. When $p(\xi, t)$ is a stationary random function of time[1], we may apply the transformation

$$S_p(\xi_1, \xi_2, \omega) = 1/\pi \int_{-\infty}^{\infty} R_p(\xi_1, \xi_2, \tau) e^{-i\omega\tau} d\tau \tag{5.3}$$

to Eq. 5.1 to obtain

$$S_{p_i}(\omega) = \int_0^L \int_0^L S_p(\xi_1, \xi_2, \omega) \varphi^i(\xi_1) \varphi^i(\xi_2) d\xi_1 d\xi_2 \tag{5.4}$$

Equation 5.4 is analogous to Eq. 5.1 in the frequency domain. The cross-spectrum of the loading $S_{p_i}(\omega)$ is conveniently normalized by the spectral density at some reference point say $S_p(0, 0, \omega)$ leaving

$$S_{p_i}(\omega) = S_p(0, 0, \omega) L^2 |J_i(\omega)|^2 \tag{5.5}$$

in which

$$|J_i(\omega)|^2 = (1/L^2) \int_0^L \int_0^L \gamma(\xi_1, \xi_2, \omega) \varphi^i(\xi_1) \varphi^i(\xi_2) d\xi_1 d\xi_2 \tag{5.6}$$

= Joint Acceptance for mode i

and

$$\gamma(\xi_1, \xi_2, \omega) = S_p(\xi_1, \xi_2, \omega)/S_p(0, 0, \omega) \tag{5.7}$$

= Cross-correlation coefficient of the load

The parameter γ is $\sqrt{}$(Coherence), as defined by Eq. 2.47 and is a measure of the space–frequency separation of the pressure spectra while $|J_i|^2$ describes the correlation between the spatial distribution of pressure for mode i and specifically represents the portion of the reference power spectrum stored in that mode.

Since we are principally interested in wind effects, we may put Eqs. 5.5–5.7 in a more useful form. First it may be argued that in a homogeneous pressure field, which is often assumed for wind turbulence, the precise locations (ξ_1, ξ_2) are not particularly important; rather, the solution may be expressed in terms of the *spatial separation*.

$$\Delta\xi = \xi_1 - \xi_2 \tag{5.8}$$

which relieves the necessity of measuring and evaluating spectra at all (ξ_1, ξ_2).

Davenport pointed out[2] that the cross-correlation coefficients are normally complex but that only the real component, which refers to spanwise or cross-wind cross-correlation, is important for line structures. Further he suggested that in the region of high correlation, γ may be represented by an exponential

$$\gamma(\xi_1, \xi_2, \omega) = \gamma(\Delta\xi, \omega) = e^{-|\Delta\xi|/L_c} = e^{-c} \tag{5.9}$$

where

$$L_c = \int_0^\infty \gamma(\Delta\xi, \omega) d\xi \tag{5.10}$$

is a correlation length termed the *scale* or dimension of the disturbance and $c = |\Delta\xi|/L_c$. Further L_c may be written as

$$L_c = V/(\bar{c}\omega) \tag{5.11}$$

where V/ω is a wave length and $\bar{c} = $ a measured constant ranging from 7 in unstable conditions to 50 in stable conditions. The former value is conservative.

These simplifications lead to the approximation for the joint acceptance

$$|J_i(\omega)|^2 = (1/L^2) \int_0^L \int_0^L e^{-|\Delta\xi|/L_c} \varphi^i(\xi_1)\varphi^i(\xi_2) d\xi_1 d\xi_2 \tag{5.12}$$

Here the factor $(1/L^2)$ is provided for convenient non-dimensionalization. (Davenport[4] uses

$$N_i = \int_0^L [\varphi^i(\xi)]^2 d\xi$$

in place of L.)

MDF SYSTEMS UNDER RANDOM LOADING 53

Once the pressure spectra are known, the spectra of the generalized coordinates are found from Eq. 3.8:

$$S_{q_i} = 1/\bar{k}_i^2 |H_i(\omega)|^2 S_{p_i}(\omega) \quad (5.13)$$

in which the generalized stiffness $\bar{k}_i$ is defined by Eq. 4.33(b) and the mechanical admittance function for mode i is generalized from Eq. 3.9 as

$$|H_i(\omega)|^2 = \frac{1}{\{[1-(\omega/\bar{\omega}_i)^2]^2 + 4\beta_i^2(\omega/\bar{\omega}_i)^2\}} \quad (5.14)$$

where β_i is defined by Eq. 4.41.

As an illustration of the application of the preceding equations, we may consider Davenport's analysis of a wind loaded suspended cable[2], without delving into the details of the wind characterization at this point. The cable is assumed to be of length 4000 ft with a sag of 270 ft. We will consider only the horizontal deflection for which the first two periods are $\bar{T}_1 = 15.0$ sec. and $\bar{T}_2 = 8.0$ sec. The corresponding values of the $\bar{k}_i$ (Eq. 4.33b) are 0.0055 and 0.0193, respectively.

For suspended cables, Davenport reduces Eq. 5.12 to an expression equivalent to

$$|J_i(\omega)|^2 = (1/L^2) 2L_c/L \int_0^L [\varphi^i(\xi)]^2 d\xi$$

$$= (1/L) 2L_c/L \int_0^1 [\varphi^i(\xi/L)]^2 d(\xi/L) \quad (5.15a)$$

in which

$$\varphi^i = \sin i\pi\xi/L \quad (5.15b)$$

The numerical values tabulated by Davenport, multiplied by an appropriate conversion factor $(\pi/2)$, are

$$|J_1(\bar{\omega}_1)|^2 = 0.3362$$

and

$$|J_2(\bar{\omega}_2)|^2 = 0.1791$$

The computation of the reference pressure $S_p(0, 0, \omega)$ for use along with $|J_i(\omega)|^2$ in Eq. 5.5 is beyond our present development but we may take the results from Davenport:

54 MDF SYSTEMS UNDER RANDOM LOADING

$$S_{p_1}(\bar{\omega}_1)\bar{\omega}_1/0.01\bar{P}^2 = 1.45 \tag{5.16a}$$

$$S_{p_2}(\bar{\omega}_2)\bar{\omega}_2/0.01\bar{P}^2 = 0.59 \tag{5.16b}$$

where $\bar{P}^2 =$ a reference pressure which is proportional to the wind velocity and density and $\bar{\omega}_1$ and $\bar{\omega}_2$ are easily calculated from $\bar{T}_1$ and $\bar{T}_2$.

Now we turn to the mean square of the dynamic pressure $\sigma_{p_i}^2$ which was computed by Davenport from Eqs. 5.16 using an equation analogous to Eq. 3.13(a) as

$$\sigma_{p_1}^2/\bar{P}^2 = 0.0454 \tag{5.17}$$

and

$$\sigma_{p_2}^2/\bar{P}^2 = 0.0344 \tag{5.18}$$

We may compute the variance of the dynamic deflection, y. We first assume that modal coupling may be neglected and express the total spectrum of the deflection as the superposition of the deflection of the individual modes as

$$S_y(\xi, \omega) = \sum_{i=1}^{n} S_{q_i}(\omega)[\varphi^i(\xi)]^2 \tag{5.19}$$

The square power of $\varphi^i(\xi)$ in Eq. 5.19 is best understood by referring to the Fourier transform of $S(\omega)$, the autocorrelation function $R(\tau)$, Eq. 2.23, which involves the product in the argument. $S_{q_i}(\omega)$ is expressed in terms of the pressure spectra $S_{p_i}(\omega)$ by Eq. 5.13 so that

$$S_y(\xi, \omega) = \sum_{i=1}^{n} (1/\bar{k}_i^2)|H_i(\omega)|^2 S_{p_i}(\omega)[\varphi^i(\xi)]^2 \tag{5.20}$$

We evaluate the variance of $S_y(\xi, \omega)$ from Eq. 2.29 as

$$\sigma_y^2 = \int_0^\infty S_y(\xi, \omega)d\omega$$

$$= \sum_{i=1}^{n} (1/\bar{k}_i^2)\sigma_{p_i}^2[\varphi^i(\xi)]^2 = \sum_{i=1}^{n} \sigma_{y_i}^2 \tag{5.21}$$

in which

$$\sigma_{p_i}^2 = \int_0^\infty |H_i(\omega)|^2 S_{p_i}(\omega) d\omega \qquad (5.22)$$

and $S_{p_i}(\omega)$ is given by Eq. 5.5.
Using an approximation we have already obtained $\sigma_{p_i}^2$ ($i=1, 2$) as Eqs. 5.17 and 5.18, so that we have from Eq. 5.15(b) and 5.21

$$\sigma_{y_1}^2 = (1/0.0055)^2 \, 0.0454 \bar{P}^2 \, \sin^2 \pi\xi/L$$

and

$$\sigma_{y_2}^2 = (1/0.0193)^2 \, 0.0344 \bar{P}^2 \, \sin^2 2\pi\xi/L$$

Then the rms values are calculated as

$$\sigma_{y_1}(L/2) = 38.74 \bar{P}$$

and

$$\sigma_{y_2}(L/2) = 9.61 \, \bar{P}$$

which check with the reference values if $\bar{P} = 0.33$, whereupon

$$\sigma_{y_1}(L/2) = 12.8 \text{ ft}$$

and

$$\sigma_{y_2}(L/4) = 3.2 \text{ ft}$$

To compute the peak deflections, the rms values are multiplied by a gust factor, which will be introduced in the next chapter.

5.2 PLANE STRUCTURES

Plane systems such as plates, slabs and membranes are widely used in engineering structures exposed to wind loading.

The framework of our analysis of such systems follows the preceding treatment of slender systems closely. The principal generalization required is the introduction of a second spatial coordinate η so that the fluctuating pressure is $p(\xi, \eta, t)$. Similar generalizations are present in the dynamic analysis where the mode shapes are now $\varphi^i(\xi, \eta)$ and the various integrals are performed using the differential surface area $dA = d\xi d\eta$.

56 MDF SYSTEMS UNDER RANDOM LOADING

The autocorrelation of the generalized loading is

$$R_{p_i}(\tau) = \int^A \int^A R_p(\xi_1, \eta_1, \xi_2, \eta_2, \tau) \varphi^i(\xi_1, \eta_1) \varphi^i(\xi_2, \eta_2) dA_1 dA_2 \quad (5.23)$$

in which $A=$ the surface area and $R_p(\xi_1, \eta_1, \xi_2, \eta_2, \tau)$ is the space–time correlation function.

Next we establish the generalized pressure spectrum

$$S_{p_i} = S_p(\xi_0, \eta_0, \omega) A^2 |J_{ii}(\omega)|^2 \quad (5.24)$$

in which the joint acceptance is

$$|J_{ii}(\omega)|^2 = 1/A^2 \int^A \int^A \gamma(\xi_1, \eta_1, \xi_2, \eta_2, \omega) \varphi^i(\xi_1, \eta_1) \varphi^i(\xi_2, \eta_2) dA_1 dA_2$$
$$(5.25a)$$

and the cross-correlation coefficient of the load

$$\gamma(\xi_1, \eta_1, \xi_2, \eta_2, \omega) = S_p(\xi_1, \eta_1, \xi_2, \eta_2, \omega)/S_p(\xi_0, \eta_0, \omega) \quad (5.25b)$$

Also the exponential approximation for the cross-correlation coefficient follows as

$$\gamma(\Delta\xi, \Delta\eta, \omega) = \gamma_\xi \gamma_\eta \quad (5.26)$$

in which

$$\gamma_\xi = e^{-|\Delta\xi|/L_{c\xi}} = e^{-C_\xi} \quad (5.27a)$$

$$\gamma_\eta = e^{-|\Delta\eta|/L_{c\eta}} = e^{-C_\eta} \quad (5.27b)$$

$$L_{c\xi} = V/(\bar{c}_\xi \omega) \quad (5.28a)$$

$$L_{c\eta} = V/(\bar{c}_\eta \omega) \quad (5.28b)$$

and the remaining terms are analogous to those in Eqs. 5.9 to 5.11.

To treat cross-correlation between modes, we have

$$S_{p_{il}} = S_p(\xi_0, \eta_0, \omega) A^2 |J_{il}(\omega)|^2 \quad (5.29)$$

in which

$$|J_{il}(\omega)|^2 = 1/A^2 \int^A \int^A \gamma \varphi^i \varphi^l dA_1 dA_2 \quad (5.30)$$

Then the corresponding spectrum of the generalized coordinates is

$$S_{q_{il}} = (1/\bar{k}_i\bar{k}_l)|H_i(\omega)||H_l(\omega)|S_{p_{il}} \quad (5.31)$$

and the displacement spectrum is

$$S_w(\xi, \eta, \omega) = \sum_{i=1}^{n} \sum_{l=1}^{n} S_{q_{il}} \varphi^i(\xi, \eta)\varphi^l(\xi, \eta) \quad (5.32)$$

If the response is mainly resonance and the frequencies well separated, the modal coupling may be neglected and only the diagonal $(i = l)$ terms retained whereby

$$S_w(\xi, \eta, \omega) = \sum_{i=1}^{n} S_{q_i}[\varphi^i(\omega)]^2$$

$$= \sum_{i=1}^{n} (1/\bar{k}_i)^2|H_i(\omega)|^2 S_{p_i}[\varphi^i(\omega)]^2 \quad (5.33)$$

in which S_{p_i} is given by Eq. 5.24 and $|H_i(\omega)|^2 \simeq 1/(4\beta_i^2)$ in the resonance region.

Using Eq. 5.24, Eq. 5.33 is conveniently rewritten in a nondimensional form

$$S_w/(S_p A^2) = \sum_{i=1}^{n} (1/\bar{k}_i)^2|H_i(\omega)|^2|J_{ii}(\omega)|^2[\varphi^i(\omega)]^2 \quad (5.34)$$

Moving one step further, still considering only the resonance contributions, we may obtain the spectrum of a particular stress quantity, say at the point $(\bar{\xi}, \bar{\eta})$, by replacing $[\varphi^i(\omega)]$ by the modal stresses S_{N_i} to write

$$S_N/(S_p A^2) = \sum_{i=1}^{n} (1/\bar{k}_i)^2|H_i(\omega)|^2|J_{ii}(\omega)|^2 S_{N_i}^2(\bar{\xi}, \bar{\eta}) \quad (5.35)$$

Finally, the variance of the displacement is written in terms of the modal contributions by substituting Eqs. 5.34 and $|H(\omega)|^2 = 1/(4\beta_i^2)$ into Eq. 3.13(a) whereupon

$$\sigma_w^2 = (\pi/4)A^2 S_p \sum_{i=1}^{n} [\bar{\omega}_i/(\bar{k}_i^2 \beta_i)]|J_{ii}(\omega)|^2[\varphi^i(\omega)]^2 \quad (5.36)$$

Similarly,

$$\sigma_N^2 = (\pi/4)A^2 S_p \sum_{i=1}^{n} [\bar{\omega}_i/(\bar{k}_i^2 \beta_i)]|J_{ii}(\omega)|^2 S_{N_i}^2(\bar{\xi}, \bar{\eta}) \quad (5.37)$$

58 MDF SYSTEMS UNDER RANDOM LOADING

The modes which participate in the response may often be identified if the mode shapes are known. Consider a simply supported rectangular plate of length a and breadth b. The mode shapes are

$$\varphi^i(\xi, \eta) = \sin\frac{j\pi\xi}{a} \sin\frac{l\pi\eta}{b} \tag{5.38}$$

For a harmonic loading

$$p(\xi, \eta, t) = p_0 \cos \omega t \tag{5.39}$$

we evaluate the generalized loading from Eq. 4.64, extended to two dimensions, as

$$\bar{p}_i = \int_{-a/2}^{a/2} \int_{-b/2}^{b/2} p_0 \cos \omega t \sin\frac{j\pi\xi}{a} \sin\frac{l\pi\eta}{b} \tag{5.40}$$

$$= (4/\pi^2)(ab/jl) \quad (j \text{ and } l \text{ odd})$$

$$= 0 \quad (j \text{ or } l \text{ even})$$

Thus only the modes j, l odd enter into the sum.

As a further simplification, suppose the pressure is coincident with one of the mode shapes i.e.,

$$p(\xi, \eta, t) = p_0 \cos \omega t \sin\frac{\bar{j}\pi\xi}{a} \sin\frac{\bar{l}\pi\eta}{b} \tag{5.41}$$

Then the plate will respond only in the mode $j = \bar{j}$, $l = \bar{l}$.

For a second illustration, we consider a circular stretched membrane which may represent a suspended roof covering an open interior. A free vibration analysis is presented elsewhere[3] and the mode shapes in polar coordinates r, θ, as shown on Fig. 5.1, are given by

$$\varphi^i(r, \theta) = B_i(\kappa r) \cos j\theta \tag{5.42a}$$

in which

$$\kappa = \sqrt{(\bar{\omega}_i^2 \mu / T_0)} \tag{5.42b}$$

μ = mass/area of membrane, T_0 = initial tension and B_i = Bessel function of the first kind.

The response is usually dominated by the first two modes $i = 0,1$ which have mode shapes

MDF SYSTEMS UNDER RANDOM LOADING 59

$$\varphi^0 = B_0(\kappa r) \qquad (5.43a)$$

which is an axisymmetric volume displacing mode and

$$\varphi^i = B_1(\kappa r) \qquad (5.43b)$$

which is an antisymmetric non-volume displacing mode and where κ is computed from Eq. 5.42(b) with the appropriate value of $\bar{\omega}_i$.
The response spectrum for the displacement Eq. 5.33 is then summed over $i = 0$ and $i = 1$.

A realistic structure, however, has openings (windows and doors) and the air movement through the openings due to the roof displacements introduces air mass inertia which tends to decrease the frequency from the fully-open case. In the limiting case when all openings are closed, no air escapes and additional stiffness is provided, raising the frequency.

Assuming that the enclosure under the roof acts like a Helmholtz resonator (the radiated wave length is much larger than the enclosure dimension), the equation of motion is[3]

$$\rho_0 l' N A_0 \ddot{\xi} + \frac{\rho_0 c^2 N^2 A_0^2}{V_0}(1 + \alpha_c)\xi = 0 \qquad (5.44a)$$

in which ξ = air displacement in hole (positive inward); N, A_0 = number and area of holes; ρ_0, V_0 = density and volume of air; l' = effective length of air mass in hole; c = speed of sound, and

$$1 + \alpha_c = \frac{\omega^2 l' V_0}{c^2 N A_0} \qquad (5.44b)$$

This type of system will be considered further in Section 6.8.2.

5.3 AXISYMMETRIC STRUCTURES

Structures which are symmetric about an axis of rotation, such as hyperbolic cooling towers, silos, storage tanks and reactor housings, are conveniently described in orthogonal curvilinear coordinates which are (usually) meridional or axial, circumferential and normal. Here, we assume that the axis of rotation is vertical. The arc length s is convenient for the meridional coordinate and the angle θ is chosen for the circumferential coordinate, as shown in Figs. 5.1 and 5.2. The equation of the meridian is written as $r = r(s)$ where r is the horizontal radius shown in Fig. 5.2.

The main computational benefit which is realized as a result of the axisymmetry is the applicability of periodic Fourier expansions in terms of θ for the dependent variables and surface loading. Consequently, the two

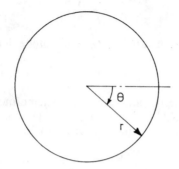

Fig. 5.1 *Plan of an axisymmetric structure*

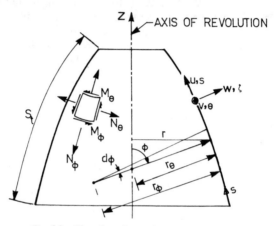

Fig. 5.2 *Elevation of an axisymmetric structure*

dimensional problem in terms of s and θ, which is described by partial differential equations, is reduced to a series of one dimensional problems, which are described by ordinary differential equations, and the total solution is found by superposition[4].

When axisymmetric structures are thin-walled, they may be considered to be thin shells of revolution for which the deformations may be described by u, v and w, the meridional, circumferential and normal displacements, as shown in Figs. 5.1 and 5.2. If the shell is cylindrical, u and w are in the vertical and radial directions. Similarly the surface loading is given by p_s, p_θ and p_ζ oriented parallel to u, v and w respectively.

The free vibration analysis of rotational shells in terms of orthogonal curvilinear coordinates is characterized by an integer number of circumferential mode shapes designated by the index j (and m). For each harmonic j, one may obtain up to n meridional mode shapes $\varphi^{(j)i}$ ($i = 1, n$) for each displacement (one for each eigenfrequency), where $n =$ the number of nodes in the discretization.

These mode shapes each contain up to n discrete ordinates, $\varphi_l^{(j)i}$ ($l = 1, n$), and may be assembled as

$$\{\varphi^{(j)i}(s)\} = \{\varphi_{u1}^{(j)i} \varphi_{v1}^{(j)i} \varphi_{w1}^{(j)i} \varphi_{u2}^{(j)i} \varphi_{v2}^{(j)i} \varphi_{w2}^{(j)i} \cdots \varphi_{un}^{(j)i} \varphi_{vn}^{(j)i} \varphi_{wn}^{(j)i}\}$$

$$(i = 1, n) \qquad (5.45)$$

Initially it is convenient to separate the mode shapes for each displacement so that

$$\{\varphi^{(j)i}(s)\}^* = \left\{\{\varphi_u^{(j)i}\}\{\varphi_v^{(j)i}\}\{\varphi_w^{(j)i}\}\right\} \qquad (5.46a)$$

in which

$$\{\varphi_u^{(j)i}\} = \{\varphi_{u1}^{(j)i} \varphi_{u2}^{(j)i} \cdots \varphi_{un}^{(j)i}\} \qquad (5.46b)$$

etc. After the modes are evaluated, Eq. 5.45 can be assembled. If the system is solved in a continuous fashion, an infinite number of meridional mode shapes for each harmonic j may be obtained.

The Fourier series expansions for the discrete dynamic displacements may be written in terms of n generalized coordinates in the form of Eq. 4.29[5] as

$$\{u\} = \sum_{j=0}^{\infty} [\Phi_u^{(j)}]\left\{\{Q^{(j)}\} \cos j\theta + \{Q^{*(j)}\} \sin j\theta\right\} \qquad (5.47)$$

$$\{v\} = \sum_{j=0}^{\infty} [\Phi_v^{(j)}]\left\{\{Q^{(j)}\} \sin j\theta - \{Q^{*(j)}\} \cos j\theta\right\} \qquad (5.48)$$

$$\{w\} = \sum_{j=0}^{\infty} [\Phi_w^{(j)}]\left\{\{Q^{(j)}\} \cos j\theta + \{Q^{*(j)}\} \sin j\theta\right\} \qquad (5.49)$$

in which the mode shapes for harmonic j are

$$[\Phi_u^{(j)}] = [\{\varphi_u^{(j)1}\}\{\varphi_u^{(j)2}\} \cdots \{\varphi_u^{(j)n}\}] \qquad (5.50a)$$

$$[\Phi_v^{(j)}] = [\{\varphi_v^{(j)1}\}\{\varphi_v^{(j)2}\} \cdots \{\varphi_v^{(j)n}\}] \qquad (5.50b)$$

$$[\Phi_w^{(j)}] = [\{\varphi_w^{(j)1}\}\{\varphi_w^{(j)2}\} \cdots \{\varphi_w^{(j)n}\}] \qquad (5.50c)$$

and the symmetric and antisymmetric generalized coordinates for harmonic j are

$$\{Q^{(j)}(t)\} = \{q_1^{(j)} q_2^{(j)} \cdots q_n^{(j)}\} \qquad (5.51a)$$

and

$$\{Q^{*(j)}(t)\} = \{q_1^{*(j)} q_2^{*(j)} \ldots q_n^{*(j)}\} \quad (5.51b)$$

In this formulation, complete Fourier series expansions for the displacements, including symmetric and antisymmetric components, are introduced to account for nonsymmetric loading. If the loading is symmetric about at least one vertical plane $\bar{\theta}$, then only the symmetric components of Eqs. 5.47–5.49 are required if $\bar{\theta}$ is taken as the origin for θ[4a].

The equations of motion follow from Eq. 4.31–4.38 as[5]

$$\ddot{q}_i^{(j)} + 2\beta_i^{(j)}\bar{\omega}_i^{(j)}\dot{q}_i^{(j)} + \bar{\omega}^{(j)2} q_i^{(j)} = \bar{p}_i^{(j)}/\bar{m}_i^{(j)}$$

$$(i = 1, n) \quad (5.52)$$

and

$$\ddot{q}_i^{*(j)} + 2\beta_i^{(j)}\bar{\omega}_i^{(j)}\dot{q}_i^{*(j)} + \bar{\omega}^{(j)2} q_i^{*(j)} = \bar{p}_i^{*(j)}/\bar{m}_i^{(j)}$$

$$(i = 1, n) \quad (5.53)$$

in which

$$\beta_i^{(j)} = \bar{c}_i^{(j)}/(2\bar{\omega}_i^{(j)}\bar{m}_i^{(j)}) \quad (5.54a)$$

$$\bar{\omega}_i^{(j)} = \sqrt{(\bar{k}_i^{(j)}/\bar{m}_i^{(j)})} \quad (5.54b)$$

$$\bar{m}_i^{(j)} = \{\varphi^{(j)i}\}^T [M^{(j)}]\{\varphi^{(j)i}\} \quad (5.54c)$$

$$\bar{k}_i^{(j)} = \{\varphi^{(j)i}\}^T [K^{(j)}]\{\varphi^{(j)i}\} \quad (5.54d)$$

$$\bar{p}_i^{(j)} = \{\varphi^{(j)i}\}^T \{P^j\} \quad (5.54e)$$

$$\bar{p}_i^{*(j)} = \{\varphi^{(j)i}\}^T \{P^{*j}\} \quad (5.54f)$$

and $[M^{(j)}]$ and $[K^{(j)}]$ are the system mass and stiffness matrices for harmonic j. Regarding the load vectors $\{P^j\}$ and $\{P^{*j}\}$, the surface loads are assumed to be represented as Fourier series

$$p_s(s, \theta, t) = \sum_{j=0}^{\infty} p_s^{(j)} \cos j\theta + p_s^{*(j)} \sin j\theta \quad (5.55a)$$

$$p_\theta(s, \theta, t) = \sum_{j=0}^{\infty} p_\theta^{(j)} \sin j\theta - p_\theta^{*(j)} \cos j\theta \quad (5.55b)$$

$$p_\zeta(s,\theta,t) = \sum_{j=0}^{\infty} p_\zeta^{(j)} \cos j\theta + p_\zeta^{*(j)} \sin j\theta \qquad (5.55c)$$

Further the loads are discretized as[5]

$$\{P^j\} = \{p_{s1}^{(j)} \cos j\theta \quad p_{\theta 1}^{(j)} \sin j\theta \quad p_{\zeta 1}^{(j)} \cos j\theta \ldots$$

$$\ldots p_{sn}^{(j)} \cos j\theta \quad p_{\theta n}^{(j)} \sin j\theta \quad p_{\zeta n}^{(j)} \cos j\theta\} \qquad (5.56)$$

and

$$\{P^{*j}\} = \{p_{s1}^{*(j)} \sin j\theta \quad p_{\theta 1}^{*(j)} \cos j\theta \quad p_{\zeta 1}^{*(j)} \sin j\theta \ldots$$

$$\ldots p_{sn}^{*(j)} \sin j\theta \quad p_{\theta n}^{*(j)} \cos j\theta \quad p_{\zeta n}^{*(j)} \sin j\theta\} \qquad (5.57)$$

for use in Eqs. 5.54(e) and (f). The solutions for the generalized coordinates follow from Eq. 4.42 with the appropriate substitutions.

An analogous set of equations for a continuous system may be written by adapting Eqs. 4.63 and 4.64:

$$\bar{m}_i^{(j)} = \int_0^{s_t} \int_0^{2\pi} \rho(s)h(s)\{[\varphi_u^{(j)i}(s)]^2 \cos^2 j\theta + [\varphi_v^{(j)i}(s)]^2 \sin^2 j\theta$$

$$+ [\varphi_w^{(j)i}(s)]^2 \cos^2 j\theta\} r(s) d\theta ds \qquad (5.58)$$

$$\bar{k}_i^{(j)} = \bar{\omega}_i^{2(j)} \bar{m}_i^{(j)} \qquad (5.59)$$

$$\bar{p}_i^{(j)} = \int_0^{s_t} \int_0^{2\pi} [p_s^{(j)}(s)\varphi_u^{(j)i}(s) \cos^2 j\theta + p_\theta^{(j)}(s)\varphi_v^{(j)i}(s) \sin^2 j\theta$$

$$+ p_\zeta^{(j)}(s)\varphi_w^{(j)i}(s) \cos^2 j\theta] r(s) d\theta ds \qquad (5.60)$$

$$\bar{p}_i^{*(j)} = \int_0^{s_t} \int_0^{2\pi} [p_s^{*(j)}(s)\varphi_u^{(j)i}(s) \sin^2 j\theta - p_\theta^{*(j)}(s)\varphi_v^{(j)i}(s) \cos^2 j\theta$$

$$+ p_\zeta^{*(j)}(s)\varphi_w^{(j)i}(s) \sin^2 j\theta] r(s) d\theta ds \qquad (5.61)$$

in which s_t = the total length of the meridian; $\rho(s)$ = mass density; and $h(s)$ = shell thickness.

Now we may write the spectra of the generalized coordinates in the form of Eq. 5.31. Possibilities are a symmetric, an antisymmetric[6] and two coupled cross-spectra which take the form

$$\begin{Bmatrix} S_{q_i^{(j)} q_l^{(m)}} \\ S_{q_i^{*(j)} q_l^{*(m)}} \\ S_{q_i^{(j)} q_l^{*(m)}} \\ S_{q_i^{*(j)} q_l^{(m)}} \end{Bmatrix} = (1/\bar{k}_i^{(j)} \bar{k}_l^{(m)}) |H_i^{(j)}| \|H_l^{(m)}| \begin{Bmatrix} S_{p_i^{(j)} p_l^{(m)}} \\ S_{p_i^{*(j)} p_l^{*(m)}} \\ S_{p_i^{(j)} p_l^{*(m)}} \\ S_{p_i^{*(j)} p_l^{(m)}} \end{Bmatrix} \quad (5.62)$$

in which $|H_i^{(j)}|$ and $|H_l^{(m)}|$ are evaluated by substituting $\bar{\omega}_i^{(j)}, \beta_i^{(j)}$ and $\bar{\omega}_l^{(m)}, \beta_l^{(m)}$, respectively, into Eq. 5.14.

We now turn to the generalized pressure spectrum as defined in Eqs. 5.24–5.25. The key term is the cross-spectrum of the pressure $S_p(s_1, \theta_1, s_2, \theta_2, \omega)$ which is generally evaluated from measurements on models or prototypes as discussed in Chapter 6. It is reasonable to expect that these measurements may be quite different for each of the three loading components p_s, p_θ and p_ζ. Since, in our applications, we will be dealing with wind forces, we will assume that S_p is established from measured values of p_ζ. Then we may adapt Eqs. 5.24–5.25 to the present geometry. Typically, for the symmetric components

$$S_{p_i^{(j)} p_i^{(m)}} = S_p(s_0, \theta_0, \omega) A^2 |J_{ii}^{(jm)}(\omega)|^2 \quad (5.63)$$

$$|J_{ii}^{(jm)}(\omega)|^2 = 1/A^2 \int_0^{s_i} \int_0^{s_i} \int_0^{2\pi} \int_0^{2\pi} \gamma(s_1, \theta_1, s_2, \theta_2, \omega)$$

$$\cdot [\varphi_w^{(j)i}(s_1) \varphi_w^{(m)i}(s_2) \cos j\theta_1 \cos m\theta_2 r(s_1) r(s_2) d\theta_1 d\theta_2 ds_1 ds_2] \quad (5.64)$$

in which

$$\gamma(s_1, \theta_1, s_2, \theta_2, \omega) = S_p(s_1, \theta_1, s_2, \theta_2, \omega)/S_p(s_0, \theta_0, \omega) \quad (5.65)$$

and A = the surface area. It is sometimes convenient to use the meridional angle φ as the variable of integration in which case $ds = r_\phi d\varphi$ where r_ϕ = the meridional radius of curvature as shown in Fig. 5.2[4b].

The remaining cross-spectra may be evaluated by replacing the trigonometric terms in Eq. 5.64 with $\sin j\theta_1 \sin m\theta_2$; $\cos j\theta_1 \sin m\theta_2$; and $\sin j\theta_1 \cos m\theta_2$, respectively, but is common to neglect the cross-correlation between symmetric and antisymmetric components.

It should also be mentioned that Eq. 5.64 contains the mode shapes as continuous function of s or Z. The conversion from the discrete forms defined in the preceding analysis may be carried out as discussed in Section 4.8 or direct numerical integration using $\{\varphi_w^{(j)i}\}$ and $\{\varphi_w^{(m)i}\}$ may be used to evaluate the integral.

A typical dynamic displacement spectrum at any location (s, θ), neglecting symmetric-antisymmetric coupling, is

$$S_w(s, \theta, \omega) = \sum_j \sum_i \sum_m \sum_l [S_{q_i^{(j)}} S_{q_l^{(m)}} \cos j\theta \cos m\theta$$

$$+ S_{q_i}^{*(j)} S_{q_l}^{*(m)} \sin j\theta \sin m\theta] \varphi_i^{(j)} \varphi_l^{(m)} \quad (5.66a)$$

and the variance is

$$\sigma_w^2 = \int_0^\infty S_w(s, \theta, \omega) d\omega \quad (5.66b)$$

Regarding the truncation limits on the summations in Eq. 5.66(a), it has been mentioned previously that a system discretized meridionally into n points will give up to n generalized coordinates. Practically, perhaps 1 to 3 meridional modes (i and l) may be sufficient for most systems. For the circumferential modes (j and m) it is often necessary to take more terms, perhaps six to eight, because of the rapid variation of the loading[6] in this direction.

Finally the peak response which is expected during a duration T may be written in a modewise fashion[6]

$$\hat{w}_i^{(j)}(s, \omega) = \tilde{w}_i^{(j)}(s, \varphi) + g_i^{(j)} \sigma_{w_i^{(j)}} \quad (5.67)$$

in which $\tilde{w}$ is the static deflection, g is a peak factor, and $\sigma_{w_i^{(j)}}$ is the standard deviation obtained from evaluating Eq. 5.66(b) for a single term of Eq. 5.66(a) with $m = j$ and $l = i$.

As an extension of Davenport's analysis[7] given in Section 3.2 which is based on a single mode, the peak factor can be expressed as

$$g_i^{(j)} = [2 \ln \bar{v}_i^{(j)} T]^{1/2} + 0.577 [2 \ln \bar{v}_i^{(j)} T]^{1/2} \quad (5.68a)$$

in which $\bar{v}_i^{(j)}$ = rate of zero crossing with positive slope of the fluctuating component of the response and is given by

$$\bar{v}_i^{(j)} = \frac{1}{2\pi} \left[\int_0^\infty \omega_i^{(j)2} S_{w_i}^{(j)}(\omega) d\omega \bigg/ \int_0^\infty S_{w_i}^{(j)}(\omega) d\omega \right]^{1/2} \quad (5.68b)$$

where $S_{w_i}^{(j)}$ is found from Eq. 5.66(a) with $m = j$ and $l = i$.

The total peak response expected during a duration T may then be obtained from the summation of the various modal contributions; the total peak factor may be from 3 to 4.5 for a tower structure.

For SDF systems such as those described in Chapter 3, the same approach is applicable if the indices (j) and i are dropped and g is evaluated using

$$\bar{v} = \bar{f} \tag{5.68c}$$

This analysis is developed further in Section 6.5.3.

In some cases, the dominant frequencies of the forcing function are far removed from those of the structure and the resonance contribution is small[8]. Then a quasistatic approximation may be employed which means $|H_i(\omega)|^2 = 1$ in Eq. 5.62. In the time domain,

$$\langle q_i^{(j)}(t) q_l^{(m)}(t) \rangle = (1/\bar{k}_i^{(j)} \bar{k}_l^{(m)}) \langle \bar{p}_i^{(j)}(t) \bar{p}_l^{(m)}(t) \rangle \tag{5.69}$$

in which the cross-correlation of the pressure is

$$\langle \bar{p}_i^{(j)}(t) \bar{p}_l^{(m)}(t) \rangle = \int_0^{s_1} \int_0^{s_2} \int_0^{2\pi} \int_0^{2\pi} \langle p_c(s_1, \theta_1, t) p_c(s_2, \theta_2, t) \rangle$$

$$\varphi_w^{(j)i}(s_1) \varphi_w^{(m)l}(s_2) \cos j\theta_1 \cos m\theta_2 r(s_1) r(s_2) d\theta_1 d\theta_2 ds_1 ds_2 \tag{5.70}$$

Similar expressions may be written for the antisymmetric components $\langle q_i^{*(j)} q_l^{*(m)} \rangle$ by substituting $\sin j\theta_1 \sin m\theta_2$ for the trigonometric terms in Eq. 5.70.

It is sometimes feasible to represent the cross-correlation of the pressure as a product $f(s)f(\theta)$ so that the integrations may be carried out separately. In nondimensional form

$$\langle \bar{p}_i^{(j)}(t) \bar{p}_l^{(m)}(t) \rangle / \{ A^2 [\sigma_p^2(s_0, \theta_0)] \} = C_s \cdot C_\theta \tag{5.71}$$

in which σ_p^2 = the variance of the pressure field at a reference point (s_0, θ_0); A = a representative area; and C_s and C_θ are evaluated from experimental data.

Finally the variance of the quasistatic displacement is

$$\sigma_w^2 = \sum_j \sum_i \sum_m \sum_l [\langle q_i^{(j)} q_l^{(m)} \rangle \cos j\theta \cos m\theta$$

$$+ \langle q_i^{*(j)} q_l^{*(m)} \rangle \sin j\theta \sin m\theta] \varphi^{(j)i} \varphi^{(m)l} \tag{5.72}$$

REFERENCES

1. Clough, R. W. and Penzien, J., *Dynamics of Structures*, McGraw-Hill, New York, 1975, pp. 451–452
2. Davenport, A. G., "The Response of Slender, Line-Like Structures to a Gusty Wind," *Inst. civ. Engrs.*, Vol. 23, 1962, pp. 389–408

3. Abu-Sitta, S. H. and Elashkar, I. D., "The Dynamic Response of Tension Roofs to Turbulent Wind," *Proc. Int. Conf. Tension Roofs*, London, April 1974
4. Gould, P. L. *Static Analysis of Shells*, Lexington Books, Lexington, Mass., 1977, pp. 107–109: (a) p. 207; (b) pp. 30–42.
5. Gould, P. L., Sen, S. K. and Suryoutomo, H., "Dynamic Analysis of Column-Supported Hyperboloidal Shells," *J. Earth. Engng. Struct. Dyn.*, Vol. 2, 1974, pp. 269–279.
6. Hashish, M. G. and Abu-Sitta, S. H., "Response of Hyperbolic Cooling Towers to Turbulent Wind," *J. Struct. Div., ASCE*, Vol. 100, No. ST5, 1974, pp. 1037–1051.
7. Davenport, A. G., "Gust Loading Factors", *J. Struct. Div., ASCE*, Vol. 93, No. ST3, June, 1967, pp. 11–34
8. Abu-Sitta, S. H. and Hashish, M. G., "Dynamic Wind Stresses in Hyperbolic Cooling Towers," *J. Struct. Div., ASCE*, Vol. 99, No. ST9, 1973, pp. 1823–1835.

6 Wind effects on structures

6.1 THE NATURE OF WIND

A physical model of a windstorm may consist of a *mean flow* determined by a large pressure system which may extend over hundreds of kilometers and superimposed *fluctuations* generated by surface roughness. The shear action of surface roughness delays the velocity of the mean flow so that it is practically zero at the surface and gradually increases with height until reaching an almost constant value, the *gradient velocity* $\bar{V}_G$ at the gradient height z_G. The region above the earth in which the flow is retarded by the surface friction is called the *boundary layer of the atmosphere* which is the location of most engineering structures.

It is therefore convenient to think of the wind velocity $V(t)$ as the sum of a time independent mean component $\bar{V}$ and a fluctuating component $\tilde{v}(t)$, i.e.

$$V(z, t) = \bar{V}(z) + \tilde{v}(z, t) \tag{6.1}$$

A power law representation for the mean velocity profiles of the form

$$\bar{V}(z) = \bar{V}_G \left[\frac{z}{z_G} \right]^\alpha \tag{6.2}$$

was suggested by Davenport[1]. In Table 6.1, the associated power law exponents, gradient height and surface drag coefficients for typical terrain conditions are given. Then in Fig. 6.1, the corresponding profiles for a uniform gradient wind velocity of 100 are shown.

The values of α are only intended to be representative of generalized surface conditions. Especially case (c), in Table 6.1, is liable to be considerably different than the situation in a particular urban setting as determined through wind tunnel measurements.

The use of the power law formula with the coefficients given in Table 6.1 may lead to high values of design wind pressure, as compared to those computed from commonly used building codes and standards, and the inclusion of these particular formulae are intended only to illustrate the general nature of the variation of $\bar{V}$ with height.

WIND EFFECTS ON STRUCTURES 69

Table 6.1 INFLUENCE OF SURFACE ROUGHNESS ON PARAMETERS RELATING TO WIND STRUCTURE NEAR THE GROUND[2]

Type of or exposure	Power law exponent α	Gradient height (ft) G	Drag coefficient K
(a) **Open terrain** with very few obstacles: e.g. open grass or farmland with few trees, hedgerows, and other barriers, etc.; prairie; tundra; shores, and low islands of inland lakes; desert	0.16	900	0.005
(b) Terrain uniformly covered with obstacles 30–50 ft in height: e.g. **residential suburbs**; small towns; **woodland**, and scrub. Small fields with bushes, trees, and hedges	0.28	1,300	0.015
(c) Terrain with large and irregular objects: e.g. **centres of large cities**; very broken country with many windbreaks of tall trees, etc.	0.40	1,700	0.050

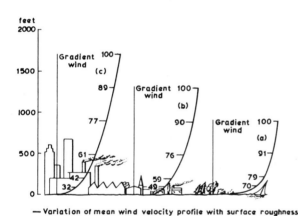

— Variation of mean wind velocity profile with surface roughness

Fig. 6.1 Wind velocity profiles (from Davenport[2])

A more recent development is a logarithmic law of the form

$$\bar{V}(z) = 2.5 V^* \ln\left(\frac{z - z_d}{z_0}\right) \quad (6.3)$$

in which

V^* = friction wind velocity which is calculated by solving Eq. 6.3 for $\bar{V}(z) = \bar{V}(z_R) = \bar{V}_{z_R}$ and $z = z_R$, where z_R is a reference height (10 m. or 30 ft) at which the standardized mean velocity $\bar{V}_{z_R}$ is measured;

z_d = zero plane displacement which is related to the average building

WIND EFFECTS ON STRUCTURES

Table 6.2 ROUGHNESS LENGTHS[3]

Exposure	Coastal (open water)	Open	Outskirts of towns, suburbs	Centre of towns	Centre of large cities
z_0 (meters)	0.005–0.01	0.03–0.10	0.20–0.30	0.35–0.45	0.60–0.80

height in the centre of large cities. It is recommended that z_d be taken as zero except in the centre of large cities where the smaller of $z_d = 20$ m. or $z_d = 0.75\,\bar{h}$ where $\bar{h}$ = the average height of the surrounding buildings may be used;
z_0 = roughness length which is related to the exposure as shown in Table 6.2.

A more detailed explanation of the logarithmic law is presented by Simiu and Lozier.

The logarithmic law is said to provide a more consistent representation of the flow field for various conditions of surface roughness and also to correspond more directly to meteorological data. Either the power law or logarithmic law may be used in what follows. In practice, the variation of $\bar{V}(z)$ is often specified by the applicable code or standard.

6.2 MEAN WIND VELOCITY RECORDS

It is desirable to take the mean velocity as a stable measure of the mean flow and also to be distinctly separable from the fluctuating component.

As a first step it is interesting to consider the power spectral density analysis of wind records of various lengths and under different conditions presented by Van der Hoven[4] and shown in Fig. 6.2. The figure also shows an extrapolation to the extreme low frequency range suggested by Davenport[2]. A more detailed study in the higher frequency range was prepared by Simiu[5].

It is clear that the energy contained in the wind is strongest in cycles measured in seconds, called *gusts*, days, weeks and longer periods. There is apparently a lack of wind energy between 10 min and 2 h. Although a period of about 10 minutes would appear to be desirable as an averaging period in order to reflect the influence of short term storms, a period of one hour has traditionally been used in Europe and Canada. A spatial distribution for Canada is shown in Fig. 6.3(a)[6]. A somewhat different concept, the *fastest-mile wind speed*, is used in the United States. This is based on observations which record the time for a mile of air to pass a fixed point and is based on the annual extreme condition as indicated in Fig. 6.3(b)[7]. The fastest mile speed, $\bar{V}_f$, may be converted to a mean speed by first computing the averaging period in seconds $t_f = 3600/\bar{V}_f$ where $\bar{V}_f$ is in m.p.h.

WIND EFFECTS ON STRUCTURES   71

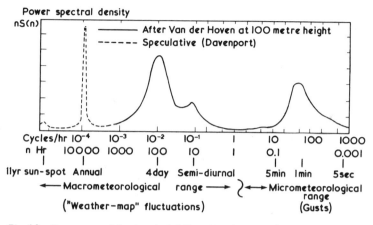

Fig. 6.2   Power spectral density of wind (from Van der Hoven[4])

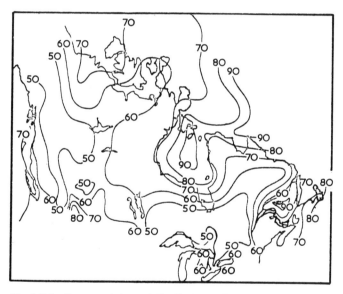

Fig. 6.3(a)   Extreme wind speed distribution in Canada. Annual probability = 1/30, hourly average speed in m.p.h. at 30 ft reference height (from Boyd[6]).

In Table 6.3 the ratio of the mean wind speed averaged over t seconds, $\bar{V}^t$, to the mean averaged over 1 h, $\bar{V}^h$, at a height 10 m. above ground in *open terrain* is shown[8].

It is possible to extend this conversion to other terrains as shown in Fig. 6.4 where ratios of maximum wind speeds are given for various exposures[9]. For example, if it is desired to convert from a wind velocity averaged over

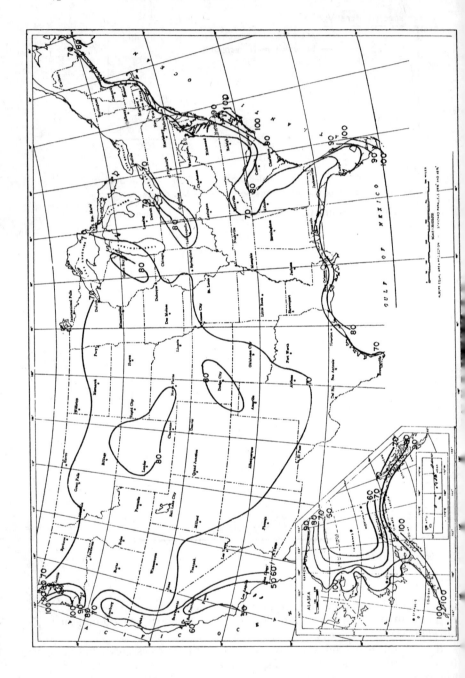

WIND EFFECTS ON STRUCTURES 73

Table 6.3 MEAN WIND SPEED RATIOS FOR OPEN TERRAIN

t_f (s)	2	5	10	30	60	100	200	500	1000	3600
$\bar{V}^t/\bar{V}^h$	1.53	1.47	1.42	1.28	1.24	1.18	1.13	1.07	1.03	1.00

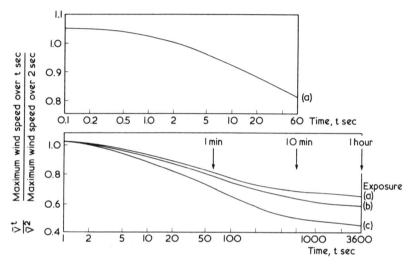

Fig. 6.4 Maximum wind speeds for three roughness categories averaged over short time intervals (from Sachs[9])

10 s in exposure (c), Table 6.1, $[\bar{V}_{(c)}^{10}]$ to the mean hourly value in the same exposure $[\bar{V}_{(c)}^{3600}]$ we would multiply it by $[\bar{V}^{3600}/\bar{V}^{10}]$ for exposure (a) $= 1/1.42$ from Table 6.3; then by

$$[\bar{V}_{(a)}^{10}]/[\bar{V}_{(c)}^{10}] = 0.92/0.87$$

opposite $t = 10$ in Fig. 6.4; and finally by

$$[\bar{V}_{(c)}^{3600}]/[\bar{V}_{(a)}^{3600}] = 0.45/0.65$$

opposite $t = 3600$ in the same figure. The mean hourly wind speed will be used for further development.

The records which are available, mean hourly or extreme annual fastest mile speed, are generally fitted with a suitable extreme value probability distribution to determine design wind speeds associated with given probabilities or risks of being exceeded. The reciprocals of these probabilities are called the *mean recurrence intervals* or *return periods* and each gives the average time interval in years between the occurrence of all winds exceeding the design value[9].

74 WIND EFFECTS ON STRUCTURES

Table 6.4 RETURN PERIODS FOR DESIGN OF STRUCTURES[7]

	Condition	R (years)
(a)	Structures having no human occupants or negligible risk to human life	25
(b)	All permanent structures except those presenting a high degree of sensitivity to wind and an unusually high degree of hazard to life and property in case of failure	50
(c)	Exceptions to (b)	100

Apparently the Fisher–Tippet Type I (Gumbel) has found favour in Europe and Canada while the Fisher–Tippet Type II (Frechet) is used in the United States[10]. For any such method, one may establish the probability that a wind velocity V is not exceeded in any year $P(V_i < V) = P(V)$ as defined in Eq. 2.4. For the United States, this has been compiled by Thom[11]. Then the return period is

$$R = 1/[1 - P(<V)] \tag{6.4}$$

The return period is only the *average* time in which V is reached or exceeded. It does not mean recurrence every R years nor does it suggest that V will certainly be exceeded only once in this period.

The probability that V is not exceeded in N consecutive years is $[P(<V)]^N$ and the probability that V is exceeded in this period is

$$P_N(>V) = 1 - [P(<V)]^N$$

$$= 1 - (1 - 1/R)^N$$

$$\simeq N/R \tag{6.5}$$

when $N \ll R$.

$P_N(>V)$ may be thought of as the *risk* which will be assumed considering human, economic and/or operational factors. If N is the lifetime of the structure, then R may be obtained from Eq. 6.5 once $P_N(>V)$ is established. Then the annual probability follows from Eq. 6.4 and the (extreme value) design wind speed is obtained from meteorological records[11] or codes[7].

For example, if the acceptable risk level $P_N(>V)$ is 5% and the economic life of the structure is 25 years, then the return period should be $25/0.05 = 500$ years. Since a return period of 30 years is common in many countries and is associated with "service loads", the ultimate design load factor may be thought of as $\text{Loads}_{R=500\text{ years}} \div \text{Loads}_{R=30\text{ years}}$.

Of course, the setting of a numerical risk is often governed by codes and standards when the public safety is involved. The ANSI Standard[7] specifies this risk directly in terms of R as in Table 6.4.

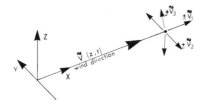

Fig. 6.5(a) Components of wind velocity

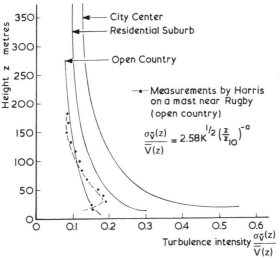

Fig. 6.5(b) RMS gust speeds (from Harris[12])

6.3 ATMOSPHERIC TURBULENCE

Following the earlier assumption that wind velocity consists of a mean component and a fluctuating component and having discussed the mean component in the previous sections, we now consider the fluctuating part $\tilde{v}(z,t)$. As shown in Fig. 6.5(a), this may be referred to a Cartesian coordinate system.

It is reasonable to represent $\tilde{v}(z,t)$ as a stationary random process and to attempt to evaluate the variance, auto-correlation and spectrum of each component and the relationships between these components.

Theoretical models show that the turbulence will decay with height, while measurements indicate that it is reasonably constant up until heights which are comparable to very tall structures. Measurements of rms gust speed by Harris[12] along with a theoretical curve are shown in Fig. 6.5(b)[13]. From this curve, the height independent approximation

WIND EFFECTS ON STRUCTURES

Table 6.5 TURBULENCE INTENSITIES

Exposure	$I(\tilde{v})$
Open	0.18
Residential suburb, woodland	0.32
City centre	0.58

$$\sigma_i = 0.11 \, \bar{V}_G \qquad (6.6)$$

appears reasonable. Note that

$$\sigma_{\tilde{v}}^2(\tilde{v}) = \sigma_{\tilde{v}_1}^2 + \sigma_{\tilde{v}_2}^2 + \sigma_{\tilde{v}_3}^3 \qquad (6.7)$$

in accordance with Fig. 6.5(a). Near the surface, $\sigma_{\tilde{v}_1}$ is much larger than the other two, indicating that the measurement of $\sigma_{\tilde{v}}$ is a strong indication of $\sigma_{\tilde{v}_1}$.

Turbulence intensity $I(\tilde{v})$ is defined as the rms gust value/mean value of the representative wind speed. It is convenient to refer this to the standardized mean velocity at 10 m above ground level, $\bar{V}_{10}$ (or $\bar{V}_{30}$ in foot units).

$$I(\tilde{v}) = \sigma_{\tilde{v}}/\bar{V}_{10} \qquad (6.8)$$

Using Eq. 6.6 together with the exposures and gradient heights from Fig. 6.1 and Table 6.1, we may evaluate $I(\tilde{v})$ (Table 6.5).

This permits $\sigma_{\tilde{v}}$ to be computed from the many available records of $\bar{V}_{10}$. Following the form of Eq. 2.29, we may rewrite Eq. 6.8 as

$$I(\tilde{v}) = \left[\int_0^\infty \frac{S_{\tilde{v}}(f)}{\bar{V}_{10}^2} df \right]^{1/2} \qquad (6.9)$$

which establishes a relationship between the turbulence intensity and the power spectral density, expressed as a function of the natural frequency.

It is now logical to write the *maximum* wind velocity in the form of Eq. 6.1 as

$$\hat{V}(z) = \bar{V}(z) + g\sigma_{\tilde{v}} \qquad (6.10)$$

in which $g = a$ *peak factor* which is a function of the dominant frequency of the turbulence and the averaging period of the mean and may be of the order 2 to 4^1. Since $\sigma_{\tilde{v}}$ is approximately constant with both height and terrain, $\hat{V}$ is the sum of a variable mean speed and a constant turbulence term.

Returning to the spectrum, attempts have been made to provide a universal expression which is representative of various sites and wind speeds over a wide frequency range. Davenport obtained a generalized spectrum from numerous measurements in the form suggested by von Kármán, and Harris modified that spectrum to the form[14]

$$\frac{fS(f)}{K\bar{V}_{10}^2} = \frac{4\bar{f}}{(2+\bar{f}^2)^{5/6}} \qquad (6.11)$$

in which K is a surface drag coefficient (Table 6.1), the reduced frequency

$$\bar{f} = fL_0/\bar{V}_{10} \qquad (6.12)$$

and L_0 is a representative scale length. Equation 6.11 is plotted in Fig. 6.6(a) for $L_0 = 1200$ m along with some theoretical and experimental test results. A more recent proposal which is expressed in terms of the friction wind velocity is given by Simiu[5] who presented two expressions

$$\frac{fS(f, z)}{(V^*)^2} = \frac{4\bar{f}^2}{(1+\bar{f}^2)^{4/3}} \qquad (6.13)$$

and

$$\frac{fS(f, z)}{(V^*)^2} = \frac{200\hat{f}}{(1+50\hat{f})^{5/3}} \qquad (6.14)$$

in which

$$\hat{f} = fz/\bar{V}(z) \qquad (6.15)$$

These expressions are plotted in Fig. 6.6(b). Equation 6.14 is seen to apply over the entire frequency range while Eq. 6.13 is applicable only in the higher range.

The coherence function γ_{xy}^2 (or the cross-correlation coefficient γ_{xy}) defined by Eq. 2.47 is useful for describing the effects of spatial separation on the structure. Following the approach of Eqs. 5.9 to 5.11 with the understanding that the modulus of S_{xy} may include both real and imaginary components, we have

$$\gamma(\xi_1, \xi_2, f) = e^{-|\Delta\xi|/L_c} \qquad (6.16)$$

in which $\xi_1 = z_1$, $\xi_2 = z_2$, and $|\Delta\xi| = |z_2 - z_1|$ for vertical separation. The length scale L_c is taken as

$$L_c(f) = \tilde{V}/(f\bar{c}) \qquad (6.17a)$$

78 WIND EFFECTS ON STRUCTURES

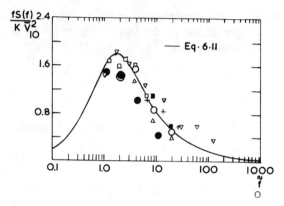

Fig. 6.6(a) Spectra of horizontal turbulence (from Macdonald[15])

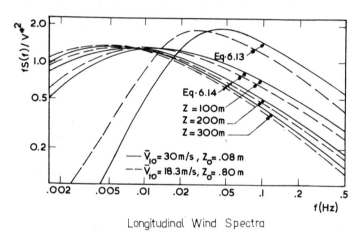

Longitudinal Wind Spectra

Fig. 6.6(b) Longitudinal wind spectra (from Simiu[5])

where

$$\tilde{V} = \tfrac{1}{2}[V(z_1) + V(z_2)] \qquad (6.17b)$$

and $\bar{c} = 6$ to 9 for wooded and open country respectively. These values are based on a plain solid wall facing the wind.

The preceding development is based on the hypothesis that the cross-correlation is based on the *separation*, not the *location*, of the two points, as explained in connection with Eq. 5.8. Another way to view that assumption is that

$$\langle \tilde{v}(x, t)\tilde{v}(x + \Delta x, t)\rangle = \langle \tilde{v}(x, t)\tilde{v}(x, t + \Delta t)\rangle \qquad (6.18)$$

where $\Delta t = \Delta x/\bar{v}$. This means that the turbulence pattern is 'frozen' such that, relative to a reference point, the flow at a distance Δx away is the same as the flow at the point after a time Δt has passed.

A more drastic assumption is homogeneous, isotropic turbulence such that

$$\sigma_{\tilde{v}_1}^2 = \sigma_{\tilde{v}_2}^2 = \sigma_{\tilde{v}_3}^2 \tag{6.19}$$

and

$$\langle \tilde{v}_1 \tilde{v}_2 \rangle = \langle \tilde{v}_1 \tilde{v}_3 \rangle = \langle \tilde{v}_2 \tilde{v}_3 \rangle = 0 \tag{6.20}$$

In this case the length scale is written from Eq. 5.10 as

$$L_1 = \int_0^\infty \gamma(\Delta\xi, f) d\xi \tag{6.21}$$

Since

$$\gamma_{x_1 x_2} = \frac{|S_{x_1 x_2}(f)|}{[S_{x_1}(f) S_{x_2}(f)]^{1/2}} \tag{6.22}$$

as given by Eq. 2.47, we may evaluate $\gamma(\Delta\xi, f)$ by applying Eq. 2.49 and 2.50. For the longitudinal length scale we take $x_1 = \tilde{v}_1(x)$ and $x_2 = \tilde{v}_1(x + \Delta x)$ whereupon

$$\int_0^\infty [S_{x_1 x_2}] d\omega = \langle \tilde{v}_1(x) \tilde{v}_1(x + \Delta x) \rangle \tag{6.23}$$

Since

$$S_{x_1} = S_{x_2} \tag{6.24}$$

$$\int_0^\infty [S_{x_1} S_{x_2}]^{1/2} d\omega = \sigma_{\tilde{v}_1}^2 \tag{6.25}$$

Then substituting Eqs. 6.23 and 6.24 into Eq. 6.22 and then into 6.21 with $\Delta\xi = \Delta x$, we have

$$L_1 = \frac{\langle \tilde{v}_1(x) \tilde{v}_1(x + \Delta x) \rangle}{\sigma_{\tilde{v}_1}^2} \tag{6.26}$$

80 WIND EFFECTS ON STRUCTURES

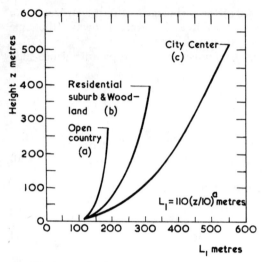

Fig. 6.7 Longitudinal length scale for turbulence (from Walsh[13])

Similar expressions may be derived for L_2 and L_3 but L_1 will be the largest. L_1 may be interpreted as the average eddy size which is moving forward at the mean wind velocity in a well-correlated manner.

A relationship for L_1 which accounts for variation in height and terrain is[13]

$$L_1 = 110 \, (z/10)^{\alpha} \quad \text{(meters)} \tag{6.27}$$

where α is given in Table 6.1. For the various exposures, Eq. 6.27 is plotted in Fig. 6.7.

6.4 WIND LOADING

Thus far, we have treated wind, both mean and turbulent, in terms of velocity. The conversion from velocity to pressure is necessary in order to provide the input to the structural model and is very sensitive since the pressure is generally proportional to the square or even higher power of the velocity. In general, this is a difficult problem and our efforts are only approximate and limited to cases which may be supported by experiments. A more complete treatment of this subject, including the nature of wind, the acquisition of pertinent data and the applications to building design, is offered by Macdonald[15] and by Simiu and Scanlan[16].

Three idealized cases are illustrated in Fig. 6.8. If a small body is inserted into the free stream, little distortion of the flow occurs and the wind load may be directly related to wind velocity. This is the case for an open-type or

WIND EFFECTS ON STRUCTURES   81

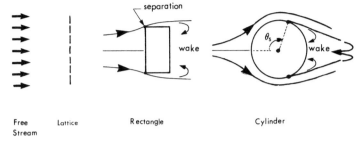

Fig. 6.8   *Action of wind on structures*

lattice structure. A large solid body, such as a rectangular building, will be obstructive or *bluff*; the flow is distorted in three geometrical directions and is separated from the building at the extreme corners, thereby enclosing the wake region. This results in the formation of circular flow patterns called eddies. These eddies or vortices are discharged at almost regular time intervals creating the phenomenon of *vortex shedding* which may produce oscillations normal to the wind direction. Also, the cross-correlation between the wake and frontal regions is generally small. The behaviour of a cylindrical body is similar except that the angle of flow separation, in the absence of sharp corners, is dependent on the surface roughness and the Reynold's number. It may be observed that for tall structures, the wind load is essentially influenced by the flow around the structure except near the top where the flow over is important. For low structures, on the other hand, the flow over the top is dominant.

The wind pressure is essentially a function of the wind velocity including the time and space variation as well as the magnitude. In order to quantify the relationships between the wind velocity and the resulting forces on the body, it is useful to introduce three basic parameters which occur frequently in this connection. First we have the dynamic pressure which provides a direct relationship between pressure and velocity

$$q(z, t) = \tfrac{1}{2}\rho V^2 \tag{6.28a}$$

in which $\rho =$ the mass density of the fluid. At the standard height of 30 ft the relationship

$$q_{30} = 0.00256\, \bar{V}_{30}^2 \tag{6.28b}$$

in which units of lb/ft² for q_{30} and $\bar{V}_{30}$ m.p.h. are routinely used. Next, we have the Reynold's number

$$R_n = Vl/\bar{v} \tag{6.29}$$

in which $l = $ a typical length of geometrically similar bodies; $\bar{v} = \bar{\mu}/\rho$

82   WIND EFFECTS ON STRUCTURES

= kinematic viscosity; and μ = viscosity of the fluid. The Reynold's number relates to the condition of flow around a bluff body by expressing a ratio between the inertial force and the frictional force.[16] Finally, we consider the Strouhal number

$$R_s = \omega l / V \qquad (6.30)$$

which relates to the frequency of shedding of vortices expressed as a function of Reynold's number.

In considering the action of wind on a structure, it is helpful to distinguish between along-wind responses (parallel to the direction of the wind) known as *drag* and lateral responses (perpendicular to the direction of the wind) known as *lift*. Also, combinations of these responses which tend to twist the structure may be important. In Sections 6.5 to 6.7, drag responses will be mainly considered while in Sections 6.8 and 6.9 lift effects will be examined.

Also, it should be recognized that there may be localized wind loadings acting on a structure which are significantly greater than the overall loads which govern the design. Particular examples are high suction wind loads near the corners of tall buildings which may break out window panes and updrafts near the base of buildings, known as *funnelling*. The later problem is related to the grouping of surrounding structures as well as to the specific structure where the funnelling occurs. Several instances have been reported where considerable local damage has occurred with no distress on the overall integrity of the structure[17].

Additionally, there may be significant wind effects which are related to neither the overall nor the local strength of the structure. These have to do with motion perception within buildings and gusts in and around built-up environments. This has been treated in some detail by Simiu and Scanlan[16] and is a subject of continuing study and research.

With respect to motion perception and resulting discomfort, accelerations of less than 0.5% g are usually imperceptible, those from 1.5% to 5% g are annoying and values beyond 15% g are intolerable. It has been suggested that the accelerations at the top of tall buildings be limited to 0.5 g except for relatively few excursions into the higher range[16]. Unfortunately it is not simple to estimate the expected accelerations from the design wind velocities.

As far as wind speed and pedestrian discomfort is concerned, velocities of 5 m/s signal the onset of discomfort; values of 10 m/s are unpleasant; while speeds of 20 m/s are dangerous[16a].

6.5   RESPONSE OF RECTANGULAR STRUCTURES

6.5.1   Pressure coefficients

We now consider a structure rectangular in plan such as a building, principally responding to the buffeting of gusts along the direction of flow. The dynamic analysis of such systems was presented in Section 5.1. If the wave length of the turbulence

$$\lambda = \bar{V}/f \qquad (6.31)$$

and λ is greater than a typical dimension of the building, it may be assumed that the flow is quasi-steady i.e., the time and space dependent variation is negligible. Then the wind pressure at the point is proportional to the dynamic pressure as expressed by the relationship

$$P(z, t) = C_D q(z, t) \qquad (6.32)$$

in which $C_D =$ a drag coefficient which depends on the geometrical shape and is generally frequency dependent.

The direct use of drag coefficients, or *pressure coefficients* as they are commonly designated, for static wind load design is quite common and is applicable to structures which are not highly sensitive to gusting action. This indicates that Eq. 6.32 in essence is replaced by

$$P(z) = C_f q(z) \qquad (6.33)$$

where C_f is a specified pressure coefficient. Compilations of pressure coefficients are given in the ANSI and similar standards for walls, roofs chimneys, tanks and towers. Here we will consider some common situations for illustrative purposes.

In applying the pressure coefficient method, it may be important to note that the basic velocity pressure $q(z)$, as specified in a design code, may be different for the structure taken as a whole "the structural frame", and for "parts and portions" of structures, with the latter case being more severe. This means that the cladding and other subsystems of the structure may be designed for higher basic pressure that the overall structure and is a reflection of possibly severe *local* wind pressures.

Thus, in Eq. 6.33, $q(z)$ might be specified as

$$q(z) = q_F(z) \quad \text{(overall structure)}$$

or

$$q(z) = q_P(z) \quad \text{(parts and portions)} \qquad (6.34)$$

A second important feature of the pressure coefficient approach as applied to *enclosed* structures is that the resulting design wind pressure may be composed of an external plus an internal component. The external component is proportional to either q_F or q_P and the internal component is proportional to an effective pressure

$$q(z) = q_M(z) \quad \text{(internal)} \qquad (6.35)$$

84 WIND EFFECTS ON STRUCTURES

Table 6.6[7] EXTERNAL PRESSURE COEFFICIENTS FOR WALLS, C_p

Location of wall	Pressure coefficient
Windward wall	0.8
Leeward wall, both height–width and height–length ratios of building ≥ 2.5	-0.6
Other buildings	-0.5
Side wall	-0.7

Table 6.7[7] EXTERNAL PRESSURE COEFFICIENTS FOR ROOFS, C_f

Gabled					θ (degrees)					
Windward slope	h/w	10–15°	20°	25°	30°	35°	40°	45°	50°	$\geq 60°$
	≤ 0.3	0.01θ*	0.2	0.25	0.3	0.35	0.4	0.45	0.5	0.01θ
	0.5	-1.0	-0.75	-0.5	-0.2	0.05	0.3	0.45	0.5	0.01θ
	0.0	-1.0	-1.0	-0.8	-0.55	-0.3	-0.05	0.2	0.45	0.01θ
	≥ 1.5	-1.0	-1.0	-1.0	-0.9	-0.6	-0.35	-0.1	0.2	0.01θ
Leeward	all	-0.7 for all	θ							
Flat	<2.5	-0.7								
	≥ 2.5	-0.8								

* Except for roofs rising from ground level ($h/w=0$), a coefficient of -1.0 shall be used when $10° < \theta < 15°$, θ = roof slope in degrees from horizontal, h = wall height at eave, w = least width of building normal to ridge.

Tabulations of q_F, q_P and q_M as a function of height, wind speed and exposures are provided in ANSI. Presumably, fluctuating as well as mean components are included in the tabulations.

Incorporating the preceding notions into Eq. 6.33 the resultant wind pressure on an enclosed structure becomes

$$P = C_p q_l - C_{pi} q_M \qquad (6.36)$$

where C_p and C_{pi} are external and internal pressure coefficients, respectively, and $q_l = q_F$ or q_P depending on whether the overall structure or a component is being designed.

In Tables 6.6 to 6.9, representative values of pressure coefficients for walls and roofs adapted from ANSI are given. The local peak coefficients (Table 6.8) may be especially severe and are to be applied outward on strips of width 0.1 w where w is defined in Table 6.7.

As an illustration we consider a gabled frame as shown in Fig. 6.9. The important parameters are $\theta = 21.80°$ and $h/w = 15/50 = 0.3$ and the pressure coefficients are recorded from the tables. For calculating C_{pi} the large door in the end wall is assumed to be the dominant opening. Also, the case of 'openings uniformly distributed' with $n=0$ is used to represent the closed structure. Now we may calculate the critical values for use in Eq. 6.36 (Table 6.10 and Fig. 6.9).

Table 6.8[7] **LOCAL PEAK EXTERNAL PRESSURE COEFFICIENTS FOR ROOFS**, C_p

Roof slope θ (degrees)	Ridges and eaves	Corners
0 to 30	−2.4	(0.1θ−5.0)
Greater than 30	−1.7	−2.0

Table 6.9[7] **INTERNAL PRESSURE COEFFICIENTS FOR BUILDINGS**, C_p

| | Openings uniformly distributed | Openings mainly in: | | |
n^*		Windward wall	Leeward wall	Side wall(s)
0 to 0.3	±0.3	(0.3 + 1.67n)	(−0.3 − n)	(−0.3 − n)
Greater than 0.3	±0.3	0.8	−0.6	−0.6

* n = ratio of open area to solid area of wall having majority of openings.

The local peak pressures will only apply to parts and portions. Also shown in the last column is a combined coefficient C_f which would be valid only if $q_F = q_P = q_M = q(z)$. In such cases, Eq. 6.33 could be used directly once a design value of $q(z)$ (or $V(z)$) were specified.

It should be noted that the determination of design wind pressures using pressure coefficients does not account for the dynamic characteristics of the structure and should be used with caution. If a gust factor is computed using the methods set out in the succeeding sections, the pressure coefficients may be useful in determining the mean pressure.

6.5.2 Pressure spectra

For structures which may have a significant sensitivity to gusting, it is often desirable to make a more sophisticated analysis.

We separate the pressure into a mean and fluctuating component

$$P(z, t) = \bar{P}(z) + p(z, t) \tag{6.37}$$

by substituting Eq. 6.1 into Eq. 6.28(a) and then into Eq. 6.32, whereby

$$\bar{P} = C_D \cdot 1/2 \rho \bar{V}^2$$

$$= C_D q(t) \tag{6.38}$$

86 WIND EFFECTS ON STRUCTURES

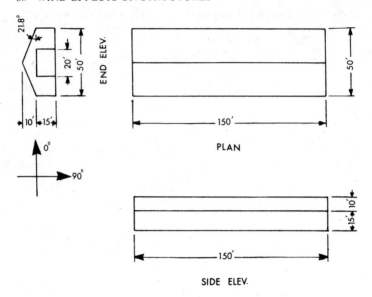

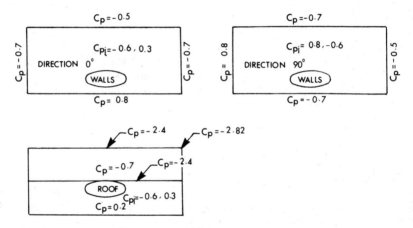

Fig. 6.9 Pressure coefficients for a gabled frame

and

$$p = C_D \rho \bar{V} \tilde{v} \tag{6.39}$$

neglecting some higher order terms[18].

Remembering that linearly proportional quantities are related by the square power in spectra, e.g., Eqs. 1.30 and 3.8 for x and P, respectively, we may write the relationship between the pressure and velocity spectra as

Table 6.10

Element	Overall structure	Parts and portions	C_f
Long walls			
Windward	$(0°)+0.8q_F-(-0.6)q_M$	$+0.8q_P-(-0.6)q_M$	$+1.4$
Leeward	$(0°)-0.5q_F-(+0.3)q_M$	$-0.5q_P-(+0.3)q_M$	-0.8
Side	$(90°)-0.7q_F-(+0.8)q_M$	$-0.7q_P-(+0.8)q_M$	-1.5
End walls			
Windward	$(90°)+0.8q_F-(-0.6)q_M$	$+0.8q_P-(-0.6)q_M$	$+1.4$
Leeward	$(90°)-0.5q_F-(+0.8)q_M$	$-0.5q_P-(+0.8)q_M$	-1.3
Side	$(0°)-0.7q_F-(+0.3)q_M$	$-0.7q_P-(+0.3)q_M$	-1.0
Roof			
Windward	$+0.2q_F-(-0.6)q_M$	$+0.2q_P-(-0.6)q_M$	$+0.8$
Leeward	$-0.7q_F-(+0.3)q_M$	$-0.7q_P-(+0.3)q_M$	-1.0
Ridges and eaves		$-2.4q_P-(+0.3)q_M$	-2.7
Corners		$-2.82q_P-(+0.3)q_M$	-3.12

$$S_p(f)=(C_D\rho \bar{V}^2)S_v(f)$$

$$=\left(\frac{2\bar{P}}{\bar{V}}\right)^2 S_v(f) \qquad (6.40)$$

Since Eq. 6.39 is an approximation and does not reflect the frequency dependence of C_D, it is appropriate to modify the relationship by introducing an *aerodynamic admittance function* χ^2 such that

$$S_p(f)=\left(\frac{2\bar{P}}{\bar{V}}\right)^2 \chi^2 S_v(f) \qquad (6.41)$$

where χ^2 reflects the frequency dependence and, in the case of a slender building, also the influence of the planform and turbulence wave length λ[18]. If the building is approximately square, $\sqrt{A}(=\sqrt{Area})$ is a convenient representative dimension for the planform.

A compilation of some studies of the aerodynamic admittance function are presented in Fig. 6.10[19]. The suggested empirical relationship

$$\chi(f)=[1+(2f\sqrt{A}/\bar{V})^{4/3}]^{-1} \qquad (6.42)$$

reflects the limits whereby $\chi \to 1$ for a low frequency, a small structure or a high mean velocity and $\chi \to 0$ for a high frequency, a large structure or a low mean velocity.

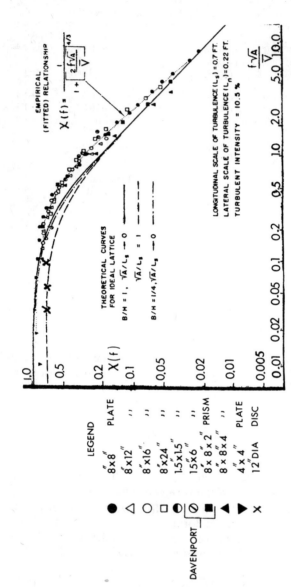

Fig. 6.10 Aerodynamic admittance function for a cantilever structure (from Vickery[19])

6.5.3 Gust factors

First, we consider the pressure spectra for a narrow plane of height $z = H$ and breadth $y = D$ inserted in a flow parallel to the x axis. The generalized pressure spectrum is given by Eq. 5.24 with $\xi = z$ and $\eta = y$. Inasmuch as the principal variation is in the z direction, it appears reasonable to define the reference pressure spectrum

$$S_p(z_1, z_2, y_1, y_2, \omega) = \gamma_z \gamma_y \sqrt{S_p(z_1) S_p(z_2)} \qquad (6.43)$$

in which

$$\gamma_z = e^{-C_z} \qquad (6.44a)$$

and

$$\gamma_y = e^{-C_y} \qquad (6.44b)$$

in accordance with Eqs. 5.7, and 5.26–5.28. Next, considering Eq. 6.41, we assume that

$$S_{\tilde{v}}(z_1) = (\bar{V}_1/\bar{V}_0)^2 S_{\tilde{v}} \qquad (6.45a)$$

$$S_{\tilde{v}}(z_2) = (\bar{V}_2/\bar{V}_0)^2 S_{\tilde{v}} \qquad (6.45b)$$

in which $S_{\tilde{v}} = S_{\tilde{v}}(z_0)$ and z_0 is a reference height. Then, in view of Eq. 6.41,

$$\sqrt{[S_p(z_1) S_p(z_2)]} = 4 S_{\tilde{v}} (P_0^2/\bar{V}_0^2) \chi^2 (\bar{V}_1 \bar{V}_2/\bar{V}_0^2) \qquad (6.46)$$

in which $P_0 = $ a reference pressure at z_0. Now we consider the generalized force spectrum, Eq. 5.24, and substitute Eqs. 6.43–6.46 into Eq. 5.25(a) and 5.24 to get

$$S_{p_i} = 4[S_{\tilde{v}}/\sigma_{\tilde{v}}^2][\sigma_{\tilde{v}}^2/\bar{V}_0^2][P_0^2 A^2]\chi^2 J_H^2 J_D^2 \qquad (6.47)$$

in which

$$J_H^2 = \int_0^1 \int_0^1 \bar{V}_1 \bar{V}_2 / \bar{V}_0^2 \varphi_i(z_1) \varphi_i(z_2) e^{-C_z} d\bar{z}_1 d\bar{z}_2 \qquad (6.48)$$

and

$$J_D^2 = \int_0^1 \int_0^1 e^{-C_y} dy_1 dy_2 \qquad (6.49)$$

90   WIND EFFECTS ON STRUCTURES

and $\bar{y}$ and $\bar{z}$ are non-dimensionalized horizontal and vertical coordinates. The exponential approximations for the coherence functions are discussed in detail in Section 6.3.

Next, it is of interest to evaluate the ratio of the variance to the static component of the response in order to derive a useful form for the gust factor. The variance of the generalized coordinate q_i may be calculated from Eq. 5.13 and 2.29 as

$$\sigma_{q_i}^2 = \int_0^\infty |H_i(\omega)|^2/\bar{k}_i^2 S_p(\omega) d\omega \qquad (6.50)$$

In the same mode i, the static component may be obtained by solving the equation of motion neglecting the inertial and damping terms. From Eq. 4.64

$$\bar{p}_i(t) = \int_0^H \varphi^i(z)p(z)Ddz$$

$$= A \int_0^1 \varphi^i(z)p(z)d\bar{z} \qquad (6.51)$$

where the breadth D converts pressure p to force/length $\bar{p}$. Then substituting into Eq. 4.37,

$$\bar{q}_i = 1/\bar{k}_i \cdot \bar{p}_i \qquad (6.52)$$

or

$$\bar{q}_i = P_0 A/\bar{k}_i I_i \qquad (6.53)$$

where

$$I_i = 1/P_0 \int_0^1 \varphi^i(z)p(z)d\bar{z} \qquad (6.54)$$

Then, from Eqs. 6.50, 6.47, 6.48, 6.53 and 6.54,

$$\sigma_{q_i}^2/\bar{q}_i^2 = 4\sigma_i^2/\bar{V}_0^2 \int_0^\infty [H_i(\omega)\chi J_H J_D/I_i]^2 \bar{S}_v d\omega \qquad (6.55)$$

where

$$\bar{S}_{\bar{i}} = S_{\bar{i}}/\sigma_{\bar{i}}^2 \qquad (6.56)$$

We may proceed with the assumption that the first mode is sufficient for cantilever type structures and that this mode is linear so that the index i may be dropped. Further, we recognize that the factor $[H(\omega)\chi J_H J_D/I]$ is a function of frequency, weighted for the net effect of the mechanical and aerodynamic admittances and the spatial correlation of the turbulence. The form

$$\sigma_q^2/\bar{q}^2 = 4\sigma_{\tilde{v}}^2/\bar{V}_0^2[B+R] \qquad (6.57a)$$

separates the integral into the *background* or quasistatic contribution B and the *resonance* contribution

$$R = SF/\eta \qquad (6.57b)$$

in which $S=$ a size reduction factor dependent on the fundamental frequency; $F=$ the gust energy ratio, also a function of frequency; and $\eta=$ a damping coefficient which considers both mechanical and aerodynamic effects.

Based on the foregoing, Davenport[2,1] has proposed a widely accepted representation of the maximum load in terms of the peak load:

$$P(z) = G\bar{P}(z) \qquad (6.58)$$

in which

$$G = 1 + g\sigma_P/\bar{P}$$

$$= 1 + gr\sqrt{(B+R)}$$

$$= \text{a gust factor} \qquad (6.59a)$$

$$r = 2\sigma(\tilde{v})/\bar{V}_0 = \text{a roughness factor} \qquad (6.59b)$$

and $g=$ a peak factor.

Figure 6.11[1] gives some representative curves which are useful for illustration. The curves are plotted in terms of the reduced frequency $\bar{f}$ as defined by Eq. 6.15(b). More complete curves are available in standard codes[7a].

As an example, we take a building 600 ft high and 100 ft in breadth normal to the wind direction. The location is suburban with $\bar{V}_{30} \simeq \bar{V}_{10} = 90$ ft/s averaged over one hour. From a dynamic analysis $\bar{f}=0.2$ Hz and β is

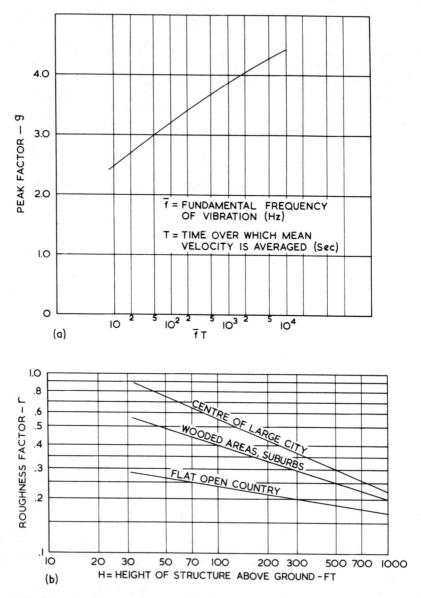

Fig. 6.11 Gust factor computation curves (from Davenport[4]); (a) peak factor, (b) roughness factor, (c) excitation caused by background turbulence, (d) size reduction factor, (e) gust energy ratio

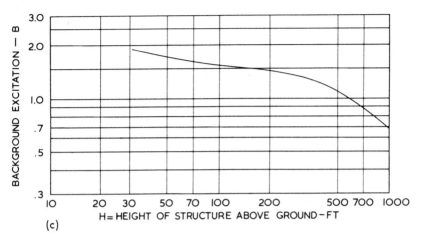

(c)

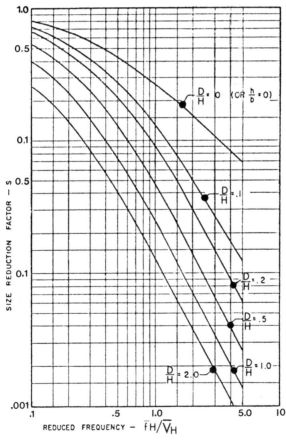

(d)

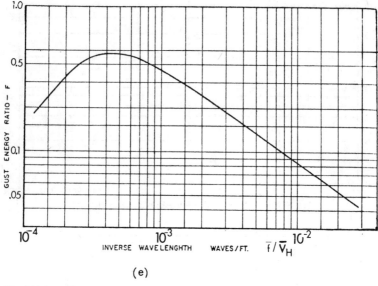

Fig. 6.11 (contd.)

estimated as 1.5%. First, from Fig. 6.11(a) with

$$\bar{f}T = 0.2 \times 3600 = 720, \quad g = 3.75$$

With $H = 600$, from Fig. 6.11(b), $r = 0.24$ and from Fig. 6.11(c), $B = 0.99$.

The reduced frequency $\tilde{f}$, originally defined in Eq. 6.12, should be scaled to the height of the structure. In accordance with Section 6.2, we refer to the ANSI Standard[7b] for the velocity scaling law. There, a coefficient $K_z(H)$ is defined such that

$$\bar{V}_H = \bar{V}_{30}\sqrt{K_z(H)} \tag{6.60}$$

$K_z(H)$ is shown in Fig. 6.12. It should be noted that exposures (a) and (c) labelled in Fig. 6.12 are interchanged from those defined in Table 6.1. For $H = 600$ ft and exposure (b), $K_z = 1.9$ and $\bar{V}_H = 90\sqrt{1.9} = 124$ ft/s. Therefore $\tilde{f}(600) = 0.2 \times 600/124 = 0.967$, and from Fig. 6.10(d) with $D/H = 0.167$, $S = 0.11$. Similarly, from Eq. 6.31, $1/\lambda_H = \bar{f}/\bar{V}_H = 0.2/124 = 1.61 \times 10^{-3}$ and $F = 0.38$ from Fig. 6.11(e). Then from Eq. 6.57(b) we compute $R = 0.11 \times 0.38/0.015 = 2.79$ and $B + R = 3.78$.

Finally, the gust factor

$$G = 1.00 + 3.75 \times 0.24\sqrt{3.78} = 2.75$$

indicating that the design wind pressure should be 2.75 times the static pressure.

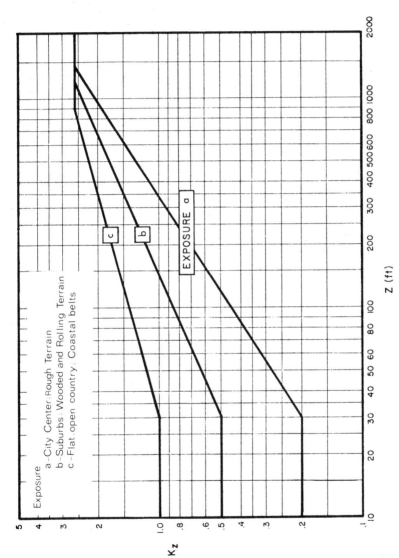

Fig. 6.12 Velocity pressure coefficients[7]

96 WIND EFFECTS ON STRUCTURES

To evaluate the static pressure $\bar{P}$ using Eq. 6.38, we estimate $C_D = 1.4$, (the sum of the windward and leeward coefficients) from Table 6.6. From Eq. 6.28(b), we compute

$$\bar{P} = 1.4 \times 0.00256(60/88)^2 \bar{V}^2$$

$$= 0.00167 \bar{V}^2 \tag{6.61}$$

where the factor (60/88) is the conversion for $\bar{V}$ from ft/s to mph. At the top of the building, $\bar{P} = 0.00167(124)^2 = 25.7$ lb/ft^2 and the peak design pressure is $2.75 \times 25.7 = 70.7$ lb/ft^2. For intermediate heights, the appropriate value of $\bar{V}$ for use in Eq. 6.61 may be found from Eq. 6.60 and Fig. 6.12[7c].

6.6 RESPONSE OF TAPERED CHIMNEYS

Slender tapered chimneys differ from other cantilever-type structures in that the diameter/height ratio is small and variable. Vickery and Kao[18] have shown that aerodynamic damping may be significant due to the slenderness of the structure and also observed that a chimney with 4% taper indicated a reduction in drag coefficient of approximately 25% from the value for a prismatic cylinder at the same subcritical Reynold's number. In comparison to Eqs. 6.38 and 6.39, they have suggested equations of the form

$$P = C_D \cdot 1/2\rho(V - \dot{w})^2 \tag{6.62}$$

Following Eq. 6.37

$$\bar{P} \simeq C_D \cdot 1/2\rho(\bar{V}^2 - 2\bar{V}\dot{w}) \tag{6.63a}$$

neglecting second order terms and

$$p = C_D \rho \bar{V} \tilde{v} \tag{6.63b}$$

where $w =$ the along wind displacement of the structure. Equation 6.62 suggests that P is proportional to the relative wind velocity $(V - \dot{w})$ rather than the actual wind velocity. The coefficient of $\dot{w}$ in Eq. 6.63(a)

$$P_a = C_D \rho \bar{V} \tag{6.64}$$

represents the *aerodynamic damping force* due to the motion of the structure. It is appropriate to convert P_a to a force per unit length $P_a D(z)$, where $D(z)$ represents the varying diameter of the stack. Then, following Eqs. 4.40, 4.35 and the generalization to continuous systems in Section 4.8,

WIND EFFECTS ON STRUCTURES 97

Table 6.11[7b] NET PRESSURE COEFFICIENTS FOR CHIMNEYS AND TANKS, C_f

		\multicolumn{3}{c}{h/d}		
Shape	Type of Surface	1	7	25
Square (wind normal to a face)	Smooth or rough	1.3	1.4	2.0
Square (wind along diagonal)	Smooth or rough	1.0	1.1	1.5
Hexanonal or octagonal $(d\sqrt{q_f} > 2.5)$	Smooth or rough	1.0	1.2	1.4
Round $(d\sqrt{q_F} > 2.5)$	Moderately smooth*	0.5	0.6	0.7
	Rough $(d'/d \simeq 0.02)$	0.7	0.8	0.9
	Very rough $(d'/d \simeq 0.08)$	0.8	1.0	1.2
$(d\sqrt{q_F} < 2.5)$		\multicolumn{3}{c}{1.2 (minimum)}		

Note: h = height of structure in feet; d = diameter or least horizontal dimension in feet; d' = depth in feet of protruding elements such as ribs and spoilers; q_F = the effective velocity pressure in psf
* Metal, timber, concrete

the aerodynamic damping factor for mode i is given by

$$\beta_{ai} = \left[\int_0^H C_D \rho D(z) \bar{V}(\varphi^i)^2 dz \right] (2\bar{\omega}_i \bar{m}_i) \qquad (6.65)$$

which should be added to the structural damping.

Pressure coefficients for chimneys as well as tanks are provided in ANSI as shown in Table 6.11. The values shown are net coefficients so that the corresponding pressure will be in the form of Eq. 6.33. This pressure is to be multiplied by the *projected area* of the structure on a vertical plane normal to the wind direction to calculate the design wind loading. According to ANSI, $P(z)$ should be at least 10 psf for the overall structure and at least 15 psf for the parts and portions of the structure.

For calculating the dynamic response the procedure is similar to that used for cantilever-type structures except that $\chi^2 \simeq 1$ because the turbulence wave length $>> D$; $J_D^2 = 1.0$ since the entire width is included in the differential strip; and J_H must include the variable diameter $D(z)$ and the variable drag coefficient. The form

$$J_H^2 = \int_0^1 \int_0^1 \gamma_1 \gamma_2 \gamma_3 \varphi^i(z_1) \varphi^i(z_2) e^{-C_z} d\bar{z}_1 d\bar{z}_2 \qquad (6.66)$$

in which

$$\gamma_1 = C_D(z_1) C_D(z_2) / C_D(z_0)^2 \qquad (6.67a)$$

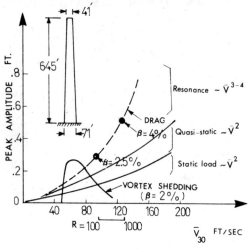

Fig. 6.13 Response of a tall chimney

$$\gamma_2 = D(z_1)D(z_2)/D(z_0)^2 \qquad (6.67b)$$

$$\gamma_3 = \bar{V}_1 \bar{V}_2 / \bar{V}_0^2 \qquad (6.67c)$$

satisfies these requirements. The expression

$$C_z = 9\tilde{f} \qquad (6.68)$$

with

$$\tilde{f} = 2\bar{f}|z_1 - z_2|/(\bar{V}_1 + \bar{V}_2) \qquad (6.69)$$

was used by Vickery and Kao[18].

The response for a tall chimney calculated by this theory is shown in Fig. 6.13. The damping ratio is increased with velocity in accordance with Eq. 6.65 and design velocities corresponding to a 100 and 1000 year return period at a particular site are indicated. The response is approximately partitioned into regions of static, quasi-static and resonance and it is apparent that the resonance region is proportional to $\bar{V}^3$ or $\bar{V}^4$ rather than the usual $\bar{V}^2$. Also shown is the vortex shedding response for $R_s = 0.2$, which is marginally significant for low frequencies in this case. The mechanics of computing the vortex shedding response will be discussed subsequently.

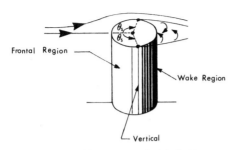

Fig. 6.14 Flow past a cylindrical object

6.7 RESPONSE OF CYLINDRICAL-TYPE TOWERS

6.7.1 Geometrical and loading characteristics

Structures which are basically axisymmetrical and relatively large such as chimneys, silos, storage tanks and cooling towers share certain aerodynamic characteristics because of similar geometric form. For such structures, the thickness/diameter ratio of the structure is an important characteristic since relatively thick shells do not suffer significant cross-sectional distortion and therefore may be treated in a similar manner as the rectangular structures and chimneys described in the previous sections. For the static pressure coefficients Table 6.11 is applicable. Here we will concentrate on structures for which the cross-sectional distortion may be significant and the use of a single C_f is inadequate. The dynamic analysis of such systems was discussed in Section 5.3.

In contrast to beams, the response of thin-walled axisymmetric structures is not only a function of the *resultant* base shear and overturning moment, but is also sensitive to the *distribution* of the forces around the circumference. The flow pattern past a cylinder is depicted in Fig. 6.14. Three independent strips may be considered: (1) A vertical strip at a given circumferential angle θ; (2) A circumferential strip in the frontal region; (3) A circumferential strip in the wake region. There is no correlation between the frontal and wake regions, and quadrature spectra, as defined by Eq. 2.43 and 2.44, are negligible. Also, cross-correlations and cross-spectra are independent of location, except along the front circumferential strips. The pressure distribution and the subsequent response of these structures is very sensitive to the boundary of the frontal and wake regions as characterized by the angle of flow separation θ_s which is strongly dependent on the Reynolds number R_n. Also the roughness of the surface is important in this regard.

6.7.2 Mean pressure distribution

In recent years, considerable effort has been directed toward the determination of the wind pressure distribution on large hyperbolic cooling

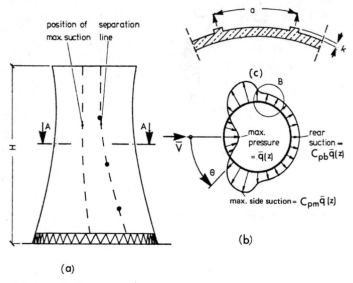

Fig. 6.15 Cooling tower under wind loading; (a) definitions, (b) section through A–A, (c) meridional wind ribs

towers and much of this work has been collected by Niemann[20]. For such structures, it is convenient to separate the circumferential pressure distribution into the three regions as shown in Fig. 6.15, maximum pressure, maximum suction and rear suction, and to take the distribution as symmetric about the wind direction axis. It should be noted that the latter assumption of symmetry may be questionable when other objects are in the vicinity of the structure, which is often the case.

Significant complications were encountered in correlating the results between model tests in wind tunnels and full scale measurements because of large Reynolds number differences. In addressing these problems two specific questions arise: (1) Which is the most consistent pressure coefficient? and (2) What is a valid measure of surface roughness? The discussion of these questions is confined to a consideration of the mean condition with a velocity profile $\bar{V}(z)$.

First, considering the various pressure coefficients identified in Fig. 6.15(b) and neglecting the static air pressure, the maximum pressure is equal to the dynamic pressure $\bar{q}(z)$ as evaluated from Eq. 6.28(a) with $V = \bar{V}(z)$. The maximum side suction is then given by $C_{pm}\bar{q}(z)$ while the rear suction is $C_{pb}\bar{q}(z)$. It has been found that the coefficient $\Delta C_p = C_{pb} - C_{pm}$, the pressure rise from peak side suction to rear suction, is independent of height. Also, for a given roughness and R_n, ΔC_p has also been found to be independent of the wind profile and therefore is a very useful basis of comparison.

Next, in considering surface roughness, it has been found that the ratio of

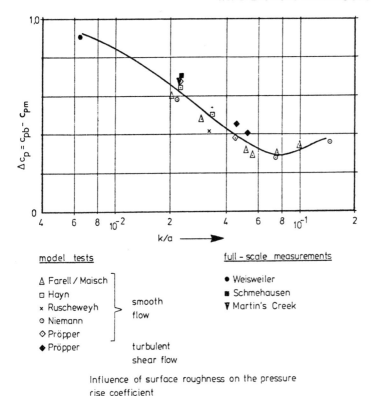

Fig. 6.15(d) Pressure coefficients as a function of surface roughness (from Niemann[20])

the height "k" to the spacing "a" of the meridional ribs, k/a, is a satisfactory roughness parameter.

A curve which gives the relationship between ΔC_p and k/a for most available model and full scale tests is shown in Fig. 6.15(d)[20] and is said to be valid for 36 to 144 ribs. Together with the use of an average value of C_{pb} (-0.4 to -0.5), Fig. 6.15(d) may be used to establish static circumferential pressure distributions for design as shown in Fig. 6.16[20]. It is common to use an internal pressure coefficient $C_{pi} = -0.5$ for such towers.

6.7.3 Fluctuating pressure distribution

We return to the analysis developed in Section 5.3 and employ the quasistatic approximation as given by Eqs. 5.69–5.72. We also neglect coupling of longitudinal modes so that the i and l subscripts are dropped.

We consider the cross-correlation of the pressure as defined by Eq. 5.70 and directly write the term $\langle p_\zeta(s_1, \theta_1, t) p_\zeta(s_2, \theta_2, t) \rangle$ in the separated form of

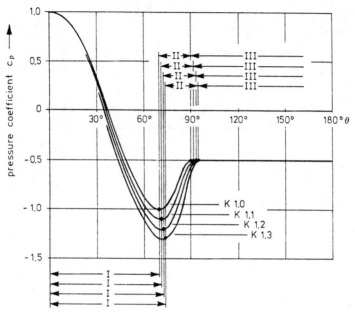

Fig. 6.16 Design circumferential pressure distributions for a cooling tower (from Niemann[20])

	k/a	I	II	III
K 1.0	3×10^{-2}	$1 - 2.0\left(\sin\dfrac{90}{70}\theta\right)^{2.267}$	$-1.0 + 0.5\left\{\sin\left[\dfrac{90}{21}(\theta-70)\right]\right\}^{2.395}$	-0.5
K 1.1	2×10^{-2}	$1 - 2.1\left(\sin\dfrac{90}{71}\theta\right)^{2.239}$	$-1.1 + 0.6\left\{\sin\left[\dfrac{90}{22}(\theta-71)\right]\right\}^{2.395}$	-0.5
K 1.2	1×2.10^{-2}	$1 - 2.2\left(\sin\dfrac{90}{72}\theta\right)^{2.205}$	$-1.2 + 0.7\left\{\sin\left[\dfrac{90}{23}(\theta-72)\right]\right\}^{2.395}$	-0.5
K 1.3	8×10^{-3}	$1 - 2.3\left(\sin\dfrac{90}{73}\theta\right)^{2.166}$	$-1.3 + 0.8\left\{\sin\left[\dfrac{90}{24}(\theta-73)\right]\right\}^{2.395}$	-0.5

Eq. 5.71, taking s_1 and $s_2 = s$ and s', and θ_1 and $\theta_2 = \theta$ and θ', respectively:

$$\langle p_\zeta(s, \theta, t) p_\zeta(s', \theta', t) \rangle / [\sigma_p(s, \theta)\sigma_p(s', \theta')] = C_s C_\theta \quad (6.70)$$

in which C_s and C_θ are non-dimensional vertical and circumferential coefficients. C_θ is different on the windward and leeward faces while C_s applies to both. Further the rms pressure is related to a reference point (z_0, θ_0) by

$$\sigma_p(s, \theta) = \sigma_p(s_0, \theta_0) f_1(s, s') f_2(\theta, \theta') \quad (6.71)$$

It is possible to approximate the rms pressure away from edge effects at the top and bottom of the tower using a two-dimensional quasi-steady theory. Provided that the pressure coefficient C_p is known, theoretical predictions showed reasonable agreement with experimental results on a model in turbulent flow[21]. In a form used by Tunstall, the rms pressure may be approximated by

$$\sigma_p(s,\theta) = \tfrac{1}{2}\rho \bar{V}^2(s)\left\{4\left[\frac{\sigma_{\tilde{v}_1}}{\bar{V}}C_p(s,\theta)\right]^2 + \left[\frac{\sigma_{\tilde{v}_2}}{\bar{V}}\frac{dC_p(s,\theta)}{d\theta}\right]^2\right\}^{1/2} \quad (6.72)$$

in which $\sigma_{\tilde{v}_2}$ is assumed to be 80% of $\sigma_{\tilde{v}_1}$.

We may also express the cross-correlation of the symmetric and antisymmetrical generalized forces defined in Eqs. 5.56 and 5.57 in a similar form:

$$\langle P^{(j)}P^{(m)}\rangle/[\sigma_p^2(s_0,\theta_0)A^2] = A_1 B_1 \quad (6.73)$$

$$\langle P^{*(j)}P^{*(m)}\rangle/[\sigma_p^2(s_0,\theta_0)A^2] = A_1 B_2 \quad (6.74)$$

in which A_1 depends on the vertical correlation which is similar to a cantilever type structure while B_1 and B_2 depend on the angle of separation. Analogous equations for a continuous system are obtained by using Eqs. 5.60 and 5.61 instead of 5.56 and 5.57 in Eqs. 6.72 to 6.74.

In Figs. 6.17[22] and 6.18, the rms pressure distributions and the cross-correlation coefficients C_s and C_θ for a hyperbolic cooling tower model at $R_n = 1.6 \times 10^3$ are shown. The meridional reference point is taken at the bottom of the tower, $s = s_0$, and the nondimensional meridional coordinate $l = s/s_0$. In general, C_s is independent of z while C_θ is strongly dependent on θ. In fact, Fig. 6.17(e) indicates a region of negative correlation. Figure 6.18(a) shows the strong dependence of the frontal correlation on θ, while Fig. 6.18(b) shows lack of correlation between the frontal and wake regions.

Computations for the generalized coefficients $A_1 B_1$ and $A_1 B_2$ for the model shown in Fig. 6.17(a) have been made and are given in Tables 6.12 and 6.13 with the substitution of S_0^4 for A^2 in Eqs. 6.73 and 6.74. For a fairly large class of towers, A_1 appears to vary only slightly while B_1 and B_2 hardly vary for a given angle of separation. The angle of separation for the model in Fig. 6.17(a) is 120° at the stated Reynold's number while that for a typical prototype tower is about 98°. Figure 6.18(c) shows the sensitivity of B_1 and B_2 to the angle of separation.

As an approximation for the quasi-steady response, Tables 6.12 and 6.13 adjusted according to Fig. 6.18(c) may be used.

Next we apply this procedure to the calculation of the response of the prototype cooling tower shown in Fig. 6.19(a)[22]. Resonance contributions are neglected for the moment, which permits the quasistatic response to be computed from Eq. 5.69, 5.70 and 5.72. The normal displacement for $\bar{q}(s) = 1.0$ is shown in Fig. 6.19(b), evaluated for j and $m = 1$ to 3 including cross

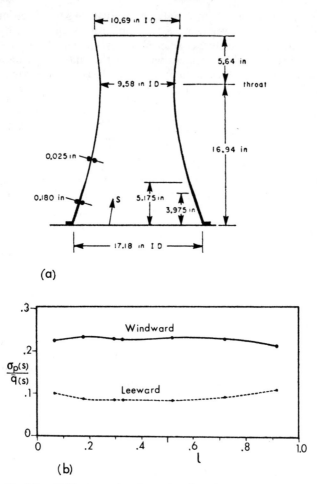

Fig. 6.17 RMS pressure distribution for a hyperbolic cooling tower model[22]. (a) Aero-elastic model dimensions, (b) rms pressure distribution along meridian, (c) rms pressure distribution along circumference at throat (d) cross-correlation coefficient along meridian, (e) cross correlation coefficient along circumference at throat level

products. The static deflection is also shown so that the total deflection takes the form of Eq. 5.67

$$w_t = w_{st} + g\sigma_w \qquad (6.75)$$

where g is the peak factor which may be 3 to 4.5 as discussed earlier. For a value of $g=4$, the corresponding meridional force and circumferential moments are shown in Figs. 6.19(c) and (d).

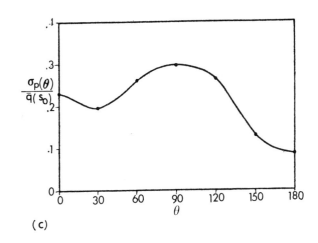

(c)

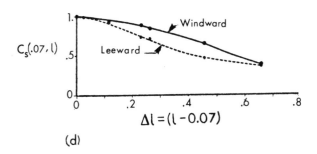

(d)

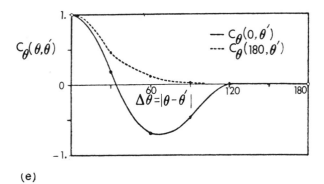

(e)

106 WIND EFFECTS ON STRUCTURES

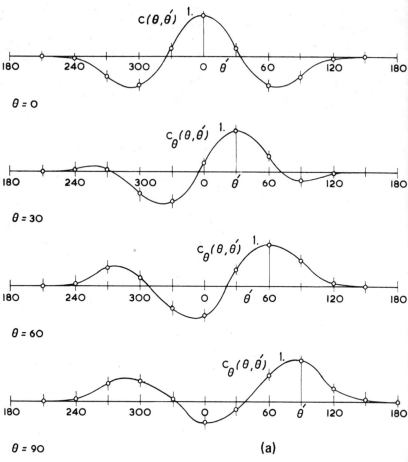

Fig. 6.18 Cross correlation coefficients for a hyperbolic cooling tower model; (a) circumferential separations at throat in windward region, (b) circumferential separation at throat in wake region

If the resonance contribution is not negligible, the time averaged quantities are insufficient to determine amplification at the natural frequencies and therefore pressure spectra must be calculated. One approach is to assume that the cross-correlation coefficients are of the form of Eqs. 5.26–5.27 and that the spectrum is directly proportional to the product of the cross-correlation coefficients and the reference pressure in the form of Eqs. 6.43 and 5.25. This leads to

$$S_p(s, \theta, s', \theta', \omega) = \gamma_s \gamma_\theta \sqrt{[S_p(s, \theta, \omega) S_p(s', \theta', \omega)]} \qquad (6.76)$$

Further details of the assumptions and justification behind Eq. 6.76 are provided elsewhere[21].

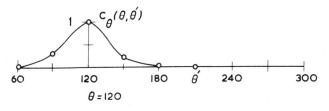

$\theta = 120$

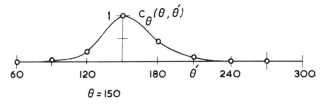

$\theta = 150$

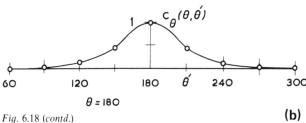

$\theta = 180$

Fig. 6.18 (contd.) **(b)**

Hashish and Abu-Sitta[21] have proposed the following numerical values for

(1) Vertical coherence

$$\gamma_s = e^{-7\tilde{f}_1} \qquad (6.77)$$

in which

$$\tilde{f}_1 = f\Delta s / \bar{V}_G \qquad (6.78)$$

for both the frontal and wake regions, and

(2) Circumferential coherence

For the wake region

$$\gamma_\theta = e^{-11\tilde{f}_2} \qquad (6.79)$$

in which

$$\tilde{f}_2 = f D(s) \Delta\theta / (2\bar{V}_G) \qquad (6.80)$$

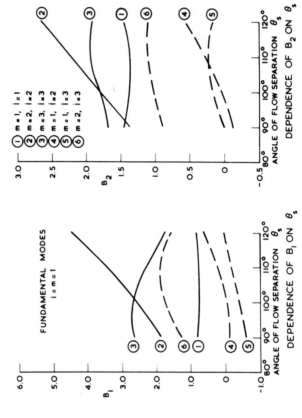

Fig. 6.18(c) Dependence of B_1 and B_2 on angle of separation.

Table 6.12 $100 A_1 B_1$

j \ m	1	2	3	4	5	6	7	8	9
1	0.934	0.811	0.050	−0.0001	−0.009	0.012	0.052	−0.012	−0.007
2		6.831	2.161	0.006	−0.0005	0.165	−0.134	0.031	0.203
3			2.105	−0.0007	−0.0017	0.059	−0.008	0.016	0.071
4				0.232	−0.039	0.014	−0.027	−0.004	0.040
5		Symmetric			0.054	−0.0008	0.004	0.003	−0.005
6						0.026	−0.027	0.001	0.027
7							0.075	−0.005	−0.044
8								0.005	0.002
9									0.058

Table 6.13 $100 A_1 B_2$

j \ m	1	2	3	4	5	6	7	8	9
1	1.655	0.738	0.204	0.003	−0.019	0.098	−0.009	−0.053	0.091
2		4.047	1.530	0.003	−0.025	0.077	0.052	−0.030	0.073
3			2.380	0.0007	−0.057	0.098	0.067	−0.065	0.106
4				0.119	0.004	0.005	−0.003	0.0005	0.005
5			Symmetric		0.095	0.004	0.003	0.013	0.018
6						0.028	0.004	−0.010	0.022
7							0.025	−0.006	0.007
8								0.014	−0.009
9									0.039

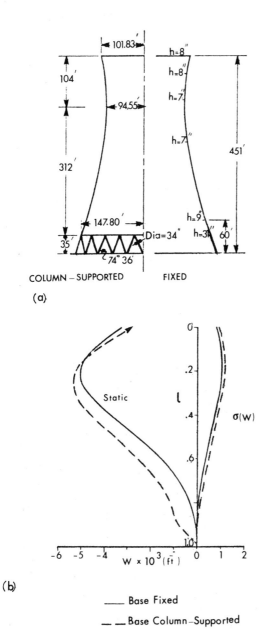

Fig. 6.19 Prototype cooling tower analysis; (a) dimensions, (b) distribution of static and rms normal deflection with height, (c) distribution of static and rms meridional force with height, (d) distribution of static and rms circumferential moment with height[21]

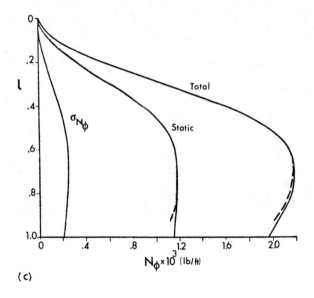

(c)

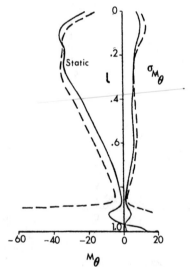

(d)

Fig. 6.19 (contd.)

WIND EFFECTS ON STRUCTURES 113

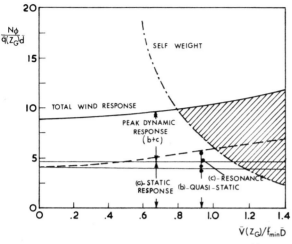

Fig. 6.20 Dynamic wind load analysis for a cooling tower[21]

For the frontal region, the circumferential coherence depends on additional factors and may be written in the form

$$\gamma_\theta = C_\theta e^{-25 f_2} \qquad (6.81)$$

in which C_θ has been defined previously and is shown in Fig. 6.18 for a model.

The experimental and computational results of the cooling tower model have been combined to produce typical expressions for the generalized force power spectra and are given elsewhere[21].

Although the use of this information for a prototype tower requires verification, the results may ultimately prove to be useful, especially in assessing a tower's vulnerability to the resonant component of the response.

An analysis was carried out in accordance with Section 5.3 for a cooling tower similar to that shown in Fig. 6.19(a) considering $j = 1$ to 5. It was observed that harmonics $j = 4$, 5 contribute mostly to the resonant response, particularly at the higher wind velocities, while $j = 1, 2, 3$ provide quasistatic response as well. This is attributable to the interesting characteristic exhibited by cylindrical and hyperboloidal shells whereby the lowest natural frequency for a given longitudinal mode *decreases* with *increasing j* to a minimum value (usually corresponding to $j = 6$ or 7 for cylinders[23] and $j = 3$ or 4 for hyperboloids[24]) and then *increases* again. The significant results are summarized in Fig. 6.20[21] where the meridional stress resultant at the lower quarter point of the tower along the windward meridian is shown as a function of gradient wind speed. The values are normalized by $\bar{q}(z_G)$ and the minimum throat diameter $\bar{D}$. The total wind response is the sum of (a) the static response; (b) the quasistatic response; and (c) the resonance response. Then the so-called "peak" dynamic response is (b) + (c). Again it is shown that the resonance may vary as the

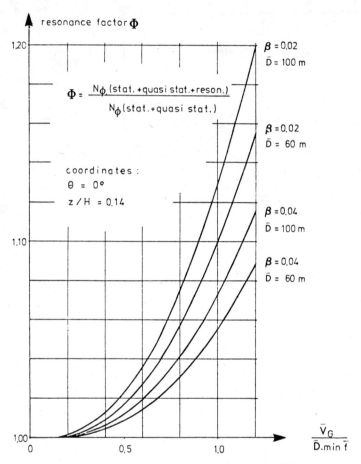

Fig. 6.21 *Resonance factors for cooling towers (from Niemann[20])*

third or fourth power of the mean velocity while the static and quasi-static are proportional to $\bar{V}^2$. Also plotted in the figure is the negative of the corresponding self-weight stress resultant. Thus, the shaded area indicates a net meridional tension which is very significant in design.

Niemann has conducted a similar study[20] in which a resonance factor

$$\Phi = N_\phi(a+b+c)/N_\phi(a+b) \tag{6.82}$$

is defined. Φ is shown for towers of typical proportions in Fig. 6.21 where the influence of damping is apparent. In this figure Z is the vertical coordinate measured from the base, and H = overall height. Also, equivalent static design loads for Middle European wind conditions for various return periods and resonance factors were proposed as shown in Fig. 6.22.

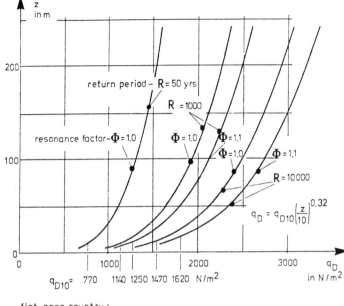

flat, open country:
exponent of mean wind profile — $\alpha = 0.16$,
gradient height — $z_G = 280$ m (900 ft).

Fig. 6.22 Static design wind loads for cooling towers (from Niemann[20])

Another approach to the problem of estimating the total dynamic response of cooling towers has recently been explored. Pressure measurements on prototype towers have been incorporated into a basically deterministic direct time history solution of the equations of motion[25,26]. Such an analysis may include the asymmetry of the wind loading which, in a time domain sense, accounts for the circumferential cross-correlation. One limitation is that the pressures were measured on only one circumferential level so that the vertical cross-correlation must be established indirectly. In Fig. 6.23 the maximum meridional force computed over a time interval of about 6 s as measured by Propper[27] is shown. Apart from the response at the beginning of the interval which may be influenced by the assumed rise time, the values of the resonance factor Φ are the same order as indicated in Fig. 6.21 for $\beta = 0.02$.

6.8 RESPONSE OF SUSPENDED ROOFS

6.8.1 Geometrical and loading characteristics

Roofs suspended by sagging cables in tension assume a geometric shape most favourable to an equilibrium position and exhibit a very high

116 WIND EFFECTS ON STRUCTURES

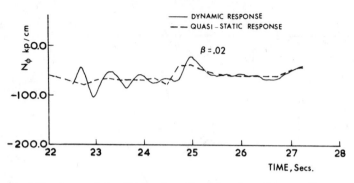

Fig. 6.23 Computed time history of meridional force in cooling tower[25]

strength-to-weight utilization of the material. However, the flexibility associated with the resulting changes in geometry may seriously affect the resistance of the structure to dynamic loading. In actual applications, the stiffness of the system may be increased by a second set of cables, often orthogonal to the first, and/or by rigid infilling material. For our purposes it will be sufficient to consider a flexible stretched membrane as a representative model. The dynamic analysis of such a system was discussed in Section 5.2.

The flow pattern over low flat roofs, as depicted in Fig. 6.24, differs from that for tall structures in that the flow is essentially parallel to the surface and a separation bubble may form at the leading edge if it tends to sag. Although the mean pressure on the upper surface is predominantly upward, the internal pressure may be downward suction depending on the number, shape, distribution and orientation of wall openings. This is indicated by Tables 6.3 and 6.9 from which the static pressure coefficients may be obtained.

6.8.2 Dynamic pressure

The dynamic pressure or fluctuating component of the wind pressure as introduced in Eq. 6.39, may be written as[28]

$$p(t) = p_w + p_r - p_i \quad (6.83)$$

in which p_w = wind turbulence pressure; p_r = acoustical radiation pressure due to roof movement; and p_i = internal pressure created by roof movement and flow of air through wall openings.

WIND EFFECTS ON STRUCTURES 117

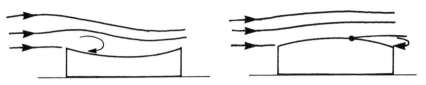

Fig. 6.24 Flow over a flat roof

Considering the turbulent wind pressure p_w, the pressure spectrum at a reference point may be taken in a form similar to Eqs. 6.43 and 6.44[28] as

$$S_{p_w}(0, 0, \omega) = \sigma_p^2(0, 0)/\omega_e \cdot e^{-\pi \rho_e^2} \qquad (6.84)$$

in which

$$\rho_e = \omega/\omega_e \qquad (6.85a)$$

$$\omega_e = \pi \bar{V}_G/(2L_1) \qquad (6.85b)$$

and L_1 is given by Eq. 6.26. This expression is compared to measured spectra on a wall parallel to the direction of flow at high velocities[29] shown in Fig. 6.25[28] and the agreement seem acceptable. In the figure, the frequency is normalized by use of the term δ^*, the boundary layer thickness which enters into the fluid mechanics considerations.

As discussed in Section 5.2, when a suspended roof is enclosed and sealed, the air beneath the roof provides an additional resistance to volume decrease resulting from vibrational modes. In effect, an added stiffness is introduced which *increases* the frequency of the roof-air system. On the other hand, the inertia of the air moving through the openings on actual structures gives rise to an additional (attached) mass which *decreases* the frequency. The frequency reduction in the fundamental (volume-displacing) mode $\bar{\omega}_0^*$ of an open structure is calculated from Eq. 5.44 and is given by[28]

$$(\bar{\omega}_0/\bar{\omega}_0^*) = 1/(1 + \beta_m) \qquad (6.86)$$

in which β_m = added mass ratio $\simeq 0.7\, \beta_c$ and β_c = cavity parameter taken as

$$\beta_c = 1.2 \rho_0 a^2/(\mu N a_0) \qquad (6.87)$$

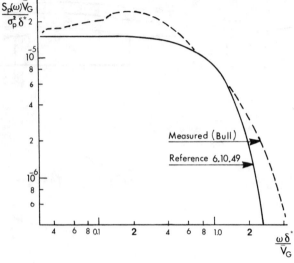

Fig. 6.25 Pressure spectra for wind acting on a wall[28]

where ρ_0 = air density; a and D = radius and diameter, respectively of membrane; μ = mass/area of membrane; N = number of wall openings; and a_0 = average radius of wall openings.

The suspension roof behaviour is governed dynamically by the wind turbulence, enclosed air underneath and, to a lesser extent, by the radiated pressure due to movement. With small openings, the frequency drops, internal pressure builds up and the total response is reduced. This effect is mainly due to the virtual disappearance of the volume-displacing mode with the lowest frequency, $\bar{\omega}_0$, which is often the only quantity estimated in practice.

With large openings, the response is greater and the resonance contribution to the fundamental mode φ^0 is considerable (0.4 to 0.7 of the total). This resonance increases as rapidly as $(\bar{V}_G/\bar{\omega}_0 D)^3$. The maximum dynamic centre deflection is of the order $0.5(\frac{1}{2}\rho_0 \bar{V}^2 D^2/Eh)$ for $(\bar{V}_G/\bar{\omega}_0 D)$ in the range 9–15, where E = Young's modulus and h = thickness of membrane. This value should be added to the static deflection[28].

Regardless of openings, the mode with one nodal diameter, φ^1, is always present and, in fact, because of it the maximum dynamic deflection is not at the centre but at about the quarter point. The nodal diameter fluctuates, giving rise to anti-symmetric components relative to the flow direction.

As an illustration of this effect, Fig. 6.26[28] shows the displacement spectrum for a suspension roof model in a suburban terrain. Results for the centre ($r/a = 0$) and the quarter point ($r/a = 0.504$) are shown for $N = 2$ and 10. For $N = 2$, frequency $\bar{\omega}_0$ is eliminated while $\bar{\omega}_1$ is practically unaffected.

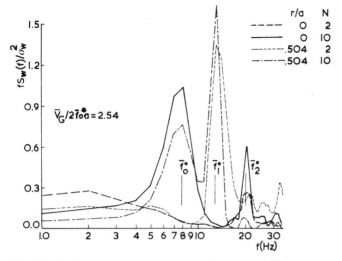

Fig. 6.26 Displacement spectrum for a suspension roof model[28]

Figure 6.26 also shows that the maximum total response occurs near the quarter point and may be up to 50% larger than the response at the centre. The dynamic deflection can exceed twice the static deflection due to wind. However, a gust factor is not easy to generalize because the mode φ^0 corresponds to a symmetrical load while φ^1 refers to an antisymmetrical load. The former depends critically on wall openings while the latter does not. The presence of the high-frequency mode, φ^2, which has a larger bending moment, may have adverse effects on stresses and crack formation in panels between the cables.

6.9 LATERAL RESPONSE OF STRUCTURES

6.9.1 Vortex shedding

The previous sections have principally dealt with structures responding to buffeting loading. There is also the possibility of the structure oscillating in a direction mainly perpendicular to the flow. Such oscillations may occur in a uniform flow without external disturbance and hence are termed *self-excited*. This is illustrated by the flow model in Fig. 6.27[15a] where a double row of vortices in the wake of a cylinder in a two dimensional flow is shown. The stable configuration of such a pattern of vortices is known as a *von Kármán Vortex Street* after the famous mechanician Theodore von Kármán. The flow pattern is affected by the Reynolds number, Eq. 6.29,

120 WIND EFFECTS ON STRUCTURES

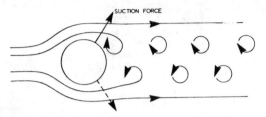

Vortex street in lee of circular cylinder.

Fig. 6.27 Karman vortex street (from Macdonald[15a])

such that for $40 < R_n < 3 \times 10^5$ the vortices detach from the obstacle and move downstream as if they were discharged *alternately* from the two sides of the cylinder. This sets up an eddying motion in the wake, which is periodic in both space and time, so-called *vortex shedding*.

The nondimensional frequency at which the vortices are shed is expressed by the Strouhal number R_s, Eq. 6.30, with $l = D =$ breadth or diameter of the structure. Considering a SDF cantilever-type structure such as a cylinder, the force per unit length which will excite lateral oscillations may be written in the form of Eq. 6.38 as

$$P = C_L \cdot 1/2 \rho \bar{V}^2 \qquad (6.88)$$

in which $C_L =$ a lift coefficient which is dependent on R_n and $\bar{V} =$ the characteristic (mean) wind velocity perpendicular to the direction of oscillation. The lift coefficient excited by vortex shedding is written in the form

$$C_L = C_{LO} e^{i\omega_s t} \qquad (6.89)$$

in which

$$\omega_s = \bar{V} R_s / D \qquad (6.90)$$

is the shedding frequency of the vortices.

For a SDF system, the *amplitude* of the oscillations associated with the steady-state solution may be written in a form similar to Eq. 1.25(a) as[30]

$$|x|/C_{LO} = \rho D / (2m\bar{\omega}^2) \cdot |H(\omega_s)| \cdot \bar{V}^2 \qquad (6.91)$$

in which $|H(\omega)|$ is given by Eq. 1.26.

In the region of resonance, $\omega_s \to \bar{\omega}$ and $H(\omega) \simeq 1/2\beta$ so that

$$|x|/C_{LO} = \beta_s \bar{V}^2 \qquad (6.92)$$

in which

$$\beta_s = \rho D/(4m\bar{\omega}^2 \beta)$$
$$= \rho D\pi/(2m\bar{\omega}^2 \bar{\delta}_s) \qquad (6.93)$$

may be thought of as a structural damping parameter which is inversely proportional to the logarithmic decrement $\bar{\delta}_s$. The extension to a MDF system is straightforward with m being replaced in Eq. 6.93 by an equivalent mass

$$m_i = \bar{m}_i/[\{\varphi^i\}^T[\varphi^i\}] \qquad (6.94)$$

as per Eq. 4.32(b) for a discrete system, or

$$m_i = \bar{m}_i / \int_0^L [\varphi^i(\xi)]^2 d\xi \qquad (6.95)$$

as per Eq. 4.63 for a continuous system.
For a wide range of R_n, $R_s \simeq 1.25^{30}$ so that for a given structure with natural frequency $\bar{\omega}$, a critical velocity

$$\bar{V}_{crit} = 0.8\bar{\omega}D \qquad (6.96)$$

may be estimated from Eq. 6.30. In reality, vortex shedding occurs over a finite frequency interval which is wider for more turbulent flow.

The oscillations induced by vortex shedding are called Aeolian harp oscillations and are the source of musical strings played by the wind, recognized since biblical times. Tall slender masts, towers, stacks and transmission lines, where $\bar{\omega}$ and D tend to be small, are prone to vortex shedding. This is illustrated in Fig. 6.13 where the critical velocity range for a tapered stack is indicated. It may be observed from Eqs. 6.91 to 6.93 that aerodynamic instabilities associated with vortex shedding may be avoided by increasing the damping parameter β_s which, in turn, is most readily accomplished by increasing the stiffness so that $\bar{\omega}$ increases, or by increasing the damping β. Also, it is possible to reduce the lift coefficient C_{LO} by altering the geometrical configuration of the structure to improve the aerodynamic shape.

In addition to the possibility of aerodynamic instability of the entire structure, it is possible for individual components such as thin-walled struts to be excited by vortex shedding. A practical procedure for treating such cases is given by Ulstrup[31].

There are situations in which it may not be practical to significantly alter the structural or geometrical characteristics in order to remedy a contemplated or actual vortex shedding problem. There are several possibilities which have been found to be satisfactory: (1) The addition of aerodynamic spoilers to avoid the generation of vortices with a predominant frequency; (2) The use of additional bracing or guying; (3) The utilization of frictional or impact damping devices; and (4) The addition of an internal lining. The specific method and device which would be most effective is strongly related to the specific type, location and function of the structure[14].

6.9.2 Galloping

The phenomenon of *galloping* is often observed in transmission lines which have been iced over. The modified cross-section is unstable because the aerodynamic forces, given by Eq. 6.88, cause a negative damping by presenting an increasing surface to the drag forces which, in turn, results in a build-up of the oscillations. These oscillations usually have a long wave length of up to one-half the entire span and exhibit amplitudes of up to 6 m as opposed to Aeolian vibrations which have shorter wave lengths and smaller amplitudes. While the Aeolian vibrations (which are much more frequent in occurrence) may be effectively controlled by dampers, such devices do not stop galloping. On some systems, ice melting is used and special framed towers are also employed. Further information is provided in Reference 32.

A recent study has indicated that a galloping instability of a single guy cable on a tall guyed tower can cause large amplitude vibrations of the guys and tower at nominal wind speeds[33].

Galloping may also occur in a building structure. Consider a nearly square building cross-section oriented toward the wind, initially at an angle of incidence $\alpha = 0$, as shown in the inset of Fig. 6.28. The lateral force exerted on the structure, Eq. 6.88, together with the drag force, Eq. 6.32, tend to introduce a rotation about the vertical axis. If

$$\frac{dC_L}{d\alpha} + C_D > 0 \qquad (6.97)$$

then the exposed surface continues to increase, energy is transmitted from the wind to the structure and large oscillations occur. Figure 6.28(a)[13] shows the regions of aerodynamic stability as a function of a normalized velocity $V_r = \bar{V}/(\bar{f}D)$ and a parameter k_s which is inversely proportional to β_s as given by Eq. 6.93. For example with $k_s = 30$, Aeolian (vortex shedding) oscillations occur for $6 < V_r < 12$ and galloping commences at $V_r = 20$.

In Fig. 6.28(b)[34] the amplitude of both types of oscillations are shown as a function of V_r. While the vortex shedding oscillations reach a limit at a critical V_r, the amplitude of the galloping oscillations increases continually with increasing wind velocity.

WIND EFFECTS ON STRUCTURES   123

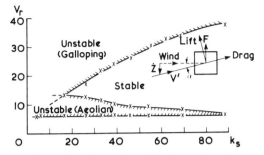

Fig. 6.28(a)   *Aerodynamic instability diagram for a long square section prism with a face normal to the wind direction (from Walshe[13])*

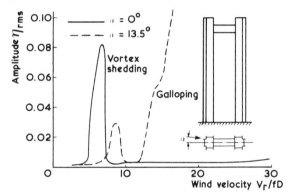

Fig. 6.28(b)   *Vortex shedding and galloping of a pair of connected sharp-edged towers (from Walshe[13])*

6.9.3 Flutter

The previously discussed types of aerodynamic oscillations are characterized by a separation of the flow so that the flow does not follow the contour of the solid body to the leeward side. Another type of self-excited oscillation which results in a combination of bending and twisting about the same axis is known as *flutter*. In streamlined bodies, flow separation does not occur and the coupled motion is called *classical flutter*. Other bodies with fairly smooth contours may exhibit similar coupled motions but also experience separation over part of the body or during part of the oscillation cycle. That is, the separation is accompanied by a reattachment in a cyclic matter. This is known as *stall flutter* and the failure of the original Tacoma Narrows Bridge is attributed to this phenomenon[30b].

The analysis of suspension bridges for coupled motion has been explored extensively since the original Tacoma Bridge failure. A basic analysis

124 WIND EFFECTS ON STRUCTURES

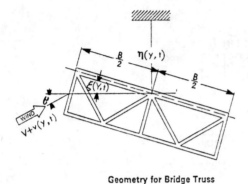

Geometry for Bridge Truss

Fig. 6.29 Cross-section of the deck of a suspension bridge (from Believeau[35], et al.)

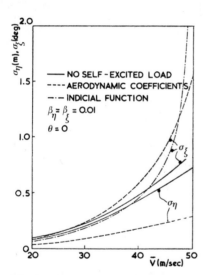

Fig. 6.30 Flutter response of a suspension bridge (from Believeau[35], et al.)

procedure is presented by Simiu and Scanlan[16b] who note that the aeroelastic stability of a bridge is controlled by the geometry of the deck, the frequencies of vibration and the mechanical damping. A representative bibliography was compiled by Beliveau, Vaicaitis and Shinozuka[35]. They also considered the analysis of the deck cross section shown in Fig. 6.29 where the longitudinal axis is designated as y. In their study, power spectral densities for turbulence are derived in a form similar to the expressions given for various types of systems in Sections 6.3 to 6.8. Of special interest, however, is the treatment of self-excited oscillations (flutter). Alternate approaches to this problem involve *indicial functions* or *aerodynamic coefficients*, both of which are dependent on experimental data. The essential results are summarized in Fig. 6.30 where the rms response is shown as a function of $\bar{V}$ for 1% damping and $\theta = 0°$. It is significant that the vertical motion is actually decreased by the inclusion of the aerodynamic loading while the twisting increases rapidly with wind velocity.

Obviously, this coupled motion can be a serious design problem. The design of the Second Tacoma Narrows Bridge incorporated deep, open stiffening trusses instead of plate girders, open trussed floor beams and streamlined rail sections. Also, the concrete deck with open slots added to the stability[30c]. Another possibility which has gained popularity in some recent bridges is to provide a streamlined cross-section so that no separation occurs. Then the problem is one of classical flutter which has a critical speed higher than that of stall flutter and which may be predicted accurately[30d].

The effect of aerodynamic forces on bridge decks is a subject of current research activity and available data and methodology are continually improving[36,37].

REFERENCES

1. Davenport, A. G., "Gust Loading Factors," *J. Struct. Div.*, *ASCE*, Vol. 93, No. ST3, June, 1967, pp. 11-34
2. Davenport, A. G., "The Application of Statistical Concepts to the Wind Loading of Structures," *Proc. Inst. civ. Engrs.*, Vol. 19, Aug. 1961, pp. 449-472
3. Simiu, E. and Lozier, D. W. "The Buffeting of Tall Structures by Strong Winds", NBS Building Science Series 74, U.S. Dept. of Commerce, Oct. 1975, pp. 3-7
4. Van der Hoven, I., "Power Spectrum of Horizontal Wind Speed in the Frequency Range from 0.0007 to 900 Cycles Per Hour," *J. Meteorology*, Vol. 14, 1967, pp. 160-164
5. Simiu, E., "Wind Spectra and Dynamic Alongwind Response," *J. Struct. Div.*, *ASCE*, Vol. 100, No. ST9, Sept. 1974, pp. 1897-1910.
6. Boyd, D. W., Climatic Supplement, National Building Code of Canada, 1970
7. American National Standard Building Code Requirements for Minimum Design Loads in Buildings and Other Structures, American National Standards Institute, ANSI A58.1, New York, N.Y., 1972 pp. 12-45: (a) pp. 46-50; (b) p. 12; (c) pp. 15-16.
8. Simiu, E., "Equivalent Static Windloads for Tall Building Design," *J. Struct. Div.*, *ASCE*, Vol. 102, No. ST4, April 1976, pp. 719-737.
9. Sachs, P., *Wind Forces in Engineering*, Pergamon Press, 1972, pp. 44-45
10. Simiu, E. and Filliben, J. J., "Statistical Analysis of Extreme Winds," NBS Technical Note 868, U.S. Dept. of Commerce, National Bureau of Standards, 1975, pp. 3-5
11. Thom, H. C. S., "New Distributions of Extreme Winds in the United States," *J. Struct. Div. ASCE*, Vol. 94, No. ST7, July, 1968, pp. 1787-1801; Vol. 95, No. ST8, Aug. 1969, p. 1769.
12. Harris, R. J., "Measurements of Wind Structure," *Symp. Wind Loading on Structures*, Bristol, June, 1972
13. Walshe, D. E. J., *Wind Excited Oscillations of Structures*, H.M.S.O., London, 1972
14. Vellozzi, J. and Cohen, E., "Dynamic Response of Tall Flexible Structures to Wind Loading," *Proc. Tech. Meeting Concerning Wind Loads on Buildings and Structures*, NBS Building Science Series 30, U.S. Dept. of Commerce, National Bureau of Standards, 1970, pp. 115-128
15. Macdonald, A. J., *Wind Loading on Buildings*, Applied Science Publishers, London, 1975; (a) p. 198

16. Simiu, E. and Scanlan, R. H., *Wind Effects on Structures*, John Wiley, New York, 1978: (a) pp. 343–385; (b) pp. 289–309
17. Koppes, W. F., *Design Wind Loads for Building Wall Elements*, in NBS Building Science Series 30, pp. 9–18
18. Vickery, B. J. and Kao, K. H., "Drag or Along-Wind Response of Slender Structures," *J. Struct. Div., ASCE*, Vol. 98, No. ST1, Jan. 1972, pp. 21–36
19. Vickery, B. J., "Load Fluctuations in Turbulent Flow," *J. Engng. Mech. Div., ASCE*, Vol. 94, No. EM1, Feb. 1968, pp. 31–46
20. Niemann, H. J., "Wind Effects on Cooling Tower Shells," *J. Struct. Div. ASCE*, Vol. 106, No. ST3, March, 1980, pp. 643–661
21. Hashish, M. G. and Abu-Sitta, S. H., "Response of Hyperbolic Cooling Towers to Turbulent Wind," *J. Struct. Div., ASCE*, Vol. 100, No. ST5, May, 1974, pp. 1037–1051.
22. Abu-Sitta, S. H. and Hashish, M. B., "Dynamic Wind Stresses in Hyperbolic Cooling Towers," *J. Struct. Div., ASCE*, Vol. 99, No. ST9, Sept. 1973, pp. 1823–1835.
23. Kraus, H., *Thin Elastic Shells*, John Wiley, New York, 1967, pp. 297–314.
24. Gould, P. L., Sen, S. K., Suryoutomo, H., "Dynamic Analysis of Column-Supported Hyperboloidal Shells," *J. Earthq. Engng. Struct. Dynamics*, Vol. 2, 1974, pp. 269–279
25. Basu, P. K. and Gould, P. L., "Cooling Towers using Measured Wind Data, *J. Struct. Div., ASCE*, Vol. 106, No. ST3, March, 1980, pp. 579–600
26. Sollenberger, N. J., Scanlan, R. H. and Billington, D. P., "Wind Loadings and Response of Cooling Towers," Preprint No. 3108, *ASCE Fall Convention*, San Francisco, Calif., Oct. 1977
27. Pröpper, H. "Zur Aerodynamischen Belastung grosser Kühltürme," *Mitteilung* Nr. 77-3, Institut für Konstruktiven Ingenierbau, Ruhr-Universität Bochum, W. Germany, Aug. 1977
28. Abu-Sitta, S. H. and Elashkar, I. D., "The Dynamic Response of Tension Roofs to Turbulent Wind," *Proc. Int. Conf. Tension Roofs*, London, April, 1974
29. Bull, M. K., "Wall-Pressure Fluctuations Associated with Subsonic Turbulent Boundary Layer Flow," *J. Fluid Mechanics*, Vol. 28, P. 4, 1967, pp. 719–754
30. Fung, Y. C., *An Introduction to the Theory of Aeroelasticity*, Dover Publications, New York, 1969, pp. 65–70: (a) p. 62; (b) pp. 62–63; (c), p. 77; (d) p. 78
31. Ulstrup, C. C. "Natural Frequencies of Axially Loaded Bridge Members", (*Technical Note J. Struct. Div. ASCE*, Vol. 104, No. ST2, Feb. 1978, pp. 357–364
32. *Motion Control*, Alcoa Conductor Products Company, Aluminium Company of America, Pittsburgh, Pa., 1976
33. Novak, M., Davenport, A. and Tanaka, H., "Vibration of Towers Due to Galloping of Iced Cables," J. Engng. Mech. Div., ASCE, Vol. 104, No. EM2, April, 1978, pp. 457–473
34. *The Modern Design of Wind Sensitive Structures*, CIRIA, London, England, 1971
35. Believeau, J. G., Vaicaitis, R. and Shinozuka, M., "Motion of Suspension Bridge Subject to Wind Loads," *J. Struct. Div., ASCE*, Vol. 103, No. ST6, June, 1977, pp. 1189–1205
36. Scanlan, H. and Lin, W.-H., "Effects of Turbulence on Bridge Flutter Derivatives," *J. Engng. Mech. Div., ASCE*, Vol. 104, No. EM4, Aug. 1978, pp. 719–733
37. Komatsu, S. and Kobayashi, H., "Experimental Identification of Aerodynamic Forces," *J. Engng. Mech. Div., ASCE*, Vol. 104, No. EM4, Aug. 1978, pp. 921–938

7 Earthquake effects on structures

7.1 THE NATURE OF EARTHQUAKES

7.1.1 General description

Earthquakes are related to global tectonic processes which are continually altering the configuration of the earth's surface. There are several geographical regions in which earthquakes occur most frequently: The Cicrum-Pacific belt around the Pacific ocean and the Alpide belt through western and central Asia are the areas where damage to man-made systems frequently occurs.

As shown in Fig. 7.1, the event is centred at a *focus* which is the point at which the rupture initiates. The point on the surface of the ground about the focus is the *epicenter* and the vertical separation of these two points is the *focal depth*. The rupture at the focus occurs because of excess straining and releases energy which propagates in various directions. The released energy is transmitted in the form of seismic waves to structures at or near the surface. Since the rupture may extend over a considerable distance, perhaps hundreds of kilometers in the case of a major earthquake, a structure may receive signals from sources all along the rupture line; hence the distance of a structure from the epicenter may be far less significant than the distance from the nearest point on the rupture surface.

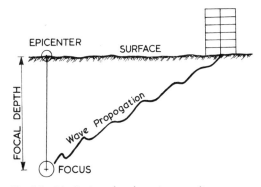

Fig. 7.1   Idealized earthquake acting on a distant structure

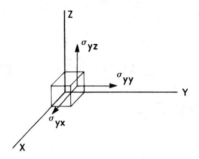

Fig. 7.2   Cartesian reference system

Earthquakes are a very frequent occurrence which are recorded every day at various seismological stations. Of these, relatively few are of interest in the engineering of structures. Significant earthquakes, in general, release a moderate to large amount of energy, are continental or near continental and have a relatively small focal depth.

Although very simplified, the foregoing description of an earthquake indicates that structures may be damaged not only by direct loss of support due to a proximate surface fault but, more likely, by base motion which results from the transmission of energy in the form of waves. The latter effect may occur at a considerable distance from the epicenter.

7.1.2   Classification systems

Common ways of classifying earthquakes are (1) the *intensity* experienced at a given location as measured by the damage to existing objects in the vicinity and the perception of motion by persons at the location; and (2) the *magnitude* of the strain energy released at the source. The intensity is measured by the modified Merchalli Intensity scale which ranges from I (Imperceptible) to XII (Great Catastrophy) with VII and above corresponding to significant structural damage. The magnitude is measured on the Richter scale which is based on recorded seismograph amplitudes. Values of greater than 5 indicate potential damage to structures. The magnitude of major earthquakes has been found to correlate with the length of the slipped fault which, in turn, reflects the size of the area affected by the shaking[1]. A magnitude of 4 corresponds to a fault length of about 1 mile while a magnitude of 8 indicates a 200 mile slip.

7.1.3   Elastic Wave Model

In order to quantify the concept of a structure responding to waves propagating from a source, it is helpful to consider the elastic wave model for an isotropic material in Cartesian coordinates as shown in Fig. 7.2.

Using indicial notation and the Einstein summation convention, the equilibrium equations are

$$\sigma_{ij,j} + f_i = 0 \qquad (i = x, y, z) \tag{7.1}$$

in which σ_{ij} = components of the stress tensor and f_i = body forces per unit volume. The corresponding kinematic law is

$$\varepsilon_{ij} = \tfrac{1}{2}(u_{i,j} + u_{j,i}) \tag{7.2}$$

in which ε_{ij} = components of the strain tensor and u_i = displacements. For an isotropic material, the constitutive law is given by

$$\sigma_{ij} = \delta_{ij}\lambda\varepsilon_{ll} + 2G\varepsilon_{ij} \tag{7.3}$$

in which ε_{ll} = volume change; E = Young's modulus; v = Poisson's ratio; $G = E/[2(1+v)]$ = shear modulus; $\lambda = \mu E/[(1+v)(1-2v)]$; and δ_{ij} = Kronecker delta.

We seek to write the equilibrium equations in terms of the displacements u_i. Substituting Eq. 7.3 into 7.1 gives

$$(\lambda\varepsilon_{ll})_{,i} + 2(G\varepsilon_{ij})_{,j} + f_i = 0 \tag{7.4}$$

since the first term remains only if $j = i$. If the material is now taken to be homogeneous as well, the material constants do not vary so that Eq. 7.4 becomes

$$\lambda(\varepsilon_{ll})_{,i} + 2G\varepsilon_{ij,j} + f_i = 0 \tag{7.5}$$

In view of Eq. 7.2, the second term of Eq. 7.5 may be written as

$$2G\varepsilon_{ij,j} = G(u_{i,j} + u_{j,i})_{,j} \tag{7.6}$$

Carrying out the differentiations,

$$u_{i,jj} = \nabla^2 u_i$$

and

$$u_{j,ij} = (u_{j,j})_{,i} = \varepsilon_{jj,i} = \varepsilon_{ll,i}$$

which permits Eq. 7.5 to be rewritten as

$$(\lambda + G)\varepsilon_{ll,i} + G\nabla^2 u_i + f_i = 0 \tag{7.7}$$

130 EARTHQUAKE EFFECTS ON STRUCTURES

The body forces f_i are assumed to be only inertial forces so that, using D'Almberts principle,

$$f_i = -\rho \ddot{u}_i \tag{7.8}$$

in which $\rho = $ mass density.

Finally we have

$$(\lambda + G)\varepsilon_{ll,i} + G\nabla^2 u_i - \rho \ddot{u}_i = 0 \tag{7.9}$$

as the general form of the elastic wave equation.

It is convenient for our purposes to consider two idealized cases of elastic wave propagation: (1) Incompressible (no volume change) and (2) Irrotational.

For the first case, $\varepsilon_{ll} = 0$ so that Eq. 7.9 becomes

$$G\nabla^2 u_i = \rho \ddot{u}_i \tag{7.10}$$

which is the elementary wave equation describing waves with a velocity of propagation

$$c_S = \sqrt{(G/\rho)} \tag{7.11}$$

The subscript S refers to the source of such waves, *shear*, which produces only rotation and no volume change.

For the second case

$$u_{u,j} - u_{j,i} = 0 \tag{7.12}$$

which is satisfied by a potential function Φ such that

$$u_i = \Phi_{,i} \tag{7.13}$$

from which

$$\varepsilon_{ll} = u_{i,i} = \Phi_{,ii} = \nabla^2 \Phi \tag{7.14}$$

Substituting Eq. 7.14 into Eq. 7.9 gives

$$(\lambda + G)(\nabla^2 \Phi)_{,i} + G\nabla^2 u_i - \rho \ddot{u}_i = 0 \tag{7.15}$$

which, in view of Eq. 7.13 becomes

$$(\lambda + G)\nabla^2 u_i = \rho \ddot{u}_i \tag{7.16}$$

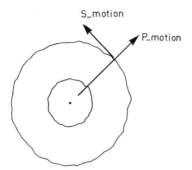

Fig. 7.3 *Wave propagation from a point source*

which is again the elementary wave equation describing waves with velocity

$$c_P = \sqrt{\{(\lambda + 2G)/\rho\}} \qquad (7.17)$$

The subscript P stands for *pressure* which is related to volume change.

If we now consider a point source in the idealized material, the wave front will consist of spheres with the motion of the P waves parallel to the direction of propagation and the motion of the S waves perpendicular to the direction of propagation as shown in Fig. 7.3. For a structure located more or less vertically above the source, the P waves would tend to introduce vertical motion while the S waves would initiate horizontal motion. Further, the S waves are polarized into vertical (SV) and horizontal (SH) waves.

Normally the geological medium is far more complicated than the idealized case which is described by the preceding equations. Particular factors which affect the wave pattern are (1) surface and interface boundaries; (2) diffraction; and (3) scattering. The free surface gives rise to *Rayleigh* (R) waves which propagate along the surface and attenuate rapidly away from the surface at a velocity slower than either P or S waves. Also, layering produces additional waves known as *Love* (L) waves.

The wave propagation from an actual earthquake source is obviously exceedingly complicated to describe in view of the foregoing. Some interesting interpretations in this regard were presented by Bolt[2] as shown in Fig. 7.4(a). The arrival times for the P, S and R waves are clearly evident. Two horizontal and one vertical component are shown in the seismograph record of the June 27, 1966 Parkfield earthquake. It is apparent that the vertical and horizontal motions are dissimilar. The vertical motion is produced by p, SV and R waves and exhibits a higher frequency content than the horizontal motion which is due to SH and L waves. However, the accelerations of the horizontal motion are considerably larger. Despite the very large maximum acceleration, this earthquake caused relatively little damage. This has been attributed to the relatively short duration. Thus we surmise that both the *peak acceleration* and the

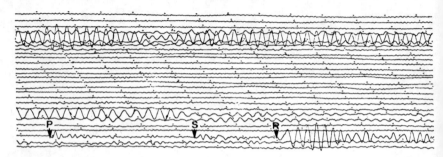

Fig. 7.4(a) Seismogram from a long-period seismograph showing the vertical component of elastic wave motion recorded at Oroville, California, during part of one day. The third trace from the bottom is from a magnitude 5 earthquake in Alaska. The time between breaks in the trace is 1 min. The maximum recorded amplitudes for P, S, and Rayleigh waves correspond to ground displacements of 1.3, 0.8, and 4.4 μ, respectively (1 $\mu = 10^{-4}$ cm) (from Bolt[2])

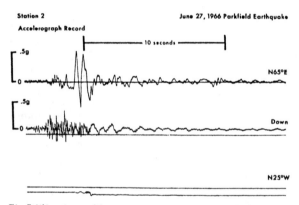

Fig. 7.4(b) A record from a strong-motion seismograph of ground acceleration near to Cholame and the San Andreas fault after the main shock of 0426 hr, June 27, 1966. (Courtesy USCGS)

duration of the shaking may be important with respect to the effect of an earthquake on a given structure. Also, recalling the concept of resonance from Chapter 1, the predominant *frequency* of the motion may be quite significant. Fagel and Liu[3] matched the fundamental frequencies of typical multistory structures with the predominant frequencies of major earthquakes as summarized in Fig. 7.5. It is interesting to observe that low-rise buildings appear to be more sensitive to resonance. Also, we see that the relatively high predominant frequency of the Parkfield earthquake, Fig. 7.4(b), would minimize resonance effects. Additionally, in considering the risk to be accepted in design, the *recurrence interval* of a significant shock is of interest.

EARTHQUAKE EFFECTS ON STRUCTURES   133

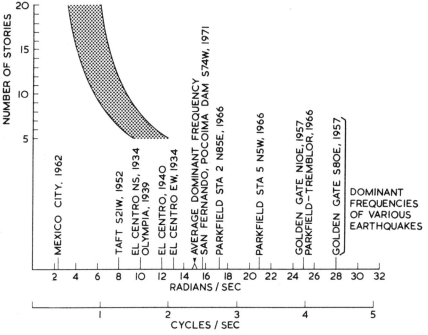

Fig. 7.5 Frequency matching of structures and earthquakes. A typical fundamental frequencies of multistory buildings and dominant frequencies of strong-motion earthquakes (from Fagel and Liu[3])

7.2 SPECTRAL ANALYSIS

7.2.1 Response spectrum

Perhaps the most descriptive representation of the influence of a given earthquake on a structure is provided by a response spectrum. We consider a SDF system similar to Fig. 1.8 subjected to a specified horizontal and vertical ground acceleration.

For each case the time history of the response may be computed from Eq. 1.44 with the base accelerations taken as $\ddot{u}_g$ and $\ddot{v}_g$ respectively. It is convenient to re-order the equation somewhat to the form of Eq. 1.37 so that the accelerations are written as $\ddot{u}_g(\tau)$ and $\ddot{v}_g(\tau)$, respectively and h becomes $h(t-\tau)$. Further, we introduce the redefined Eq. 1.35(b) into Eq. 1.44, neglect the difference between $\bar{\omega}$ and $\bar{\omega}_d$, so that $1 - \beta^2 = 1$ and take the lower limit as 0 since $t < 0$ is of no interest.

Finally we have

$$u(t, \bar{\omega}, \beta) = 1/\bar{\omega} \int_0^t \ddot{u}_g(\tau) \hat{h}(t - \tau) d\tau \qquad (7.18a)$$

$$v(t, \bar{\omega}, \beta) = 1/\bar{\omega} \int_0^t \ddot{v}_g(\tau)\hat{h}(t-\tau)d\tau \tag{7.18b}$$

in which

$$\hat{h}(t-\tau) = e^{-\beta\bar{\omega}(t-\tau)} \sin \bar{\omega}(t-\tau) \tag{7.19}$$

The next step, which is crucial, is to remove the time dependence by considering only the *maximum* values of the integrals which occur in the time range of Eqs. 7.18 at some time t_m:

$$S_{Vu}(t_m, \bar{\omega}, \beta) = \left[\int_0^t \ddot{u}_g(\tau)\hat{h}(t-\tau)d\tau \right]_{max} \tag{7.20a}$$

and

$$S_{Vv}(t_m, \bar{\omega}, \beta) = \left[\int_0^t \ddot{v}_g(\tau)\hat{h}(t-\tau)d\tau \right]_{max} \tag{7.20b}$$

The precise time t_m for which the maximum value of the integral is produced is not explicitly contained in Eqs. 7.20. This time may, in fact, be different for S_{Vu} and S_{Vv}. The ensuing maxima are, however, functions of the particular $\bar{\omega}$ and β which are chosen prior to the integration.

The maximum displacements found from Eqs. 7.18 are

$$u_{max} = (1/\bar{\omega})S_{Vu} \tag{7.21a}$$

and

$$v_{max} = (1/\bar{\omega})S_{Vv} \tag{7.21b}$$

It is apparent from the analogy to undamped simple harmonic motion ($\dot{x} = \bar{\omega}x$) that S_{V_u} and S_{V_v} have the dimension of velocity, hence the subscript V, and are called the maximum pseudo-velocity responses to the ground acceleration. The term 'pseudo' is technically necessary because damping is involved but is often omitted for simplicity.

The preceding development is for a given SDF system with a specified $\bar{\omega}$ and β. However the concept may be generalized by (1) holding β constant and computing a set of $S_{Vu}(\bar{\omega})$ and $S_{Vv}(\bar{\omega})$ for a range of $\bar{\omega}$; and (2) repeating (1) for a range of β. The results of such a computation for the N–S horizontal

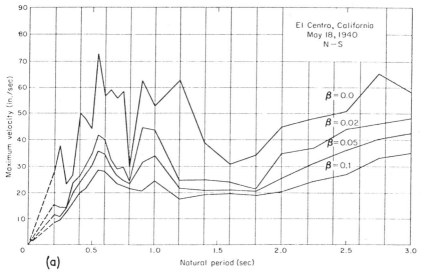

Fig. 7.6(a) Velocity response spectrum for El Centro Earthquake of May 18, 1940 (N–S) (from Clough [4])

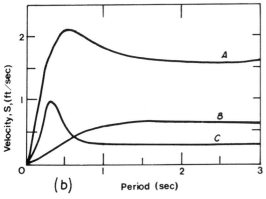

Fig. 7.6(b) Idealized undamped velocity spectrum curves that illustrate the effect of magnitude and distance. Curve A, 25 mi from centre of large earthquake; curve B, 70 mi from centre of large shock; curve C, 8 mi from centre of small (M = 5.3) shock.

component of the May 18, 1940 El Centro earthquake are shown in Fig. 7.6(a)[4]. In this graph, the abscissa was chosen as the natural period $\bar{T} = 2\pi/\bar{\omega}$. It is apparent that such a plot records the maximum (pseudo) velocity for a large family of SDF systems within the range of $\bar{T}$ and β considered. Moreover, recalling the discussion of Section 4.6, we recognize that a wide range of MDF systems are also included in such a family of curves when these systems are described in generalized coordinates. Thus,

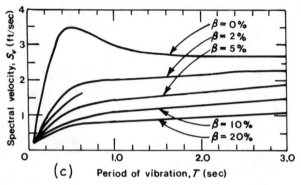

Fig. 7.6(c)   *Average velocity response spectrums*, 1940 El Centro intensity (*from Clough*[4])

one family of curves, such as given in Fig. 7.6(a), represents the *response* of a wide *spectrum* of possible structures at a given site due to a specified ground motion; hence, the term *response spectrum* is appropriate.

It is obvious that the same event will produce non-identical effects at different recording stations. This is illustrated in Fig. 7.6(b). Quite often 'average' spectra incorporating records from several sites are used for design purposes, as illustrated by Fig. 7.6(c).

In the following development, the u and v designations are dropped when the treatment for both cases is identical. It is often assumed that the horizontal and vertical effects are uncorrelated but there are some exceptions as discussed by Rosenbluth and Contreras[5]. Although there is ordinarily correlation between the two horizontal components, this correlation is minimal if principal directions are used. Penzien and Watabe[6] have found that such directions are (1) toward the epicenter and (2) normal to the epicenter. However, standardized earthquake records are given with respect to North–South and East–West directions corresponding to instrument orientations. Thus, it may be necessary to consider correlation between the effects caused by different horizontal components of the design earthquake. This is discussed further by Rosenbluth and Contreras[5]. Here, we will confine our treatment to a single component.

Using Eqs. 7.21, we may evaluate the displacement response spectrum as

$$S_D = (1/\bar{\omega})S_V \tag{7.22}$$

as shown in Fig. 7.7.

Similarly, the acceleration response spectra

$$S_A = \bar{\omega} S_V = \bar{\omega}^2 S_D \tag{7.23}$$

may be plotted as shown in Fig. 7.8. Equation 7.23 should be developed further however, since we must differentiate between the relative acceleration and the absolute acceleration which includes the ground acceleration, as discussed in Section 1.5. If we are primarily interested in the

EARTHQUAKE EFFECTS ON STRUCTURES   137

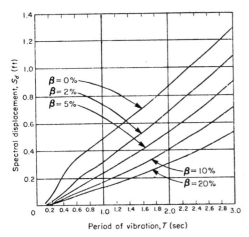

Fig. 7.7   Average displacement response spectrum, 1940 El Centro intensity (from Clough[4])

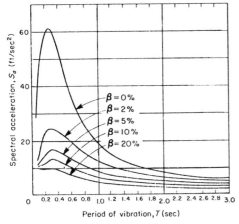

Fig. 7.8   Average acceleration response spectrums, 1940 El Centro intensity.

maximum force in the system

$$F = kS_D = m\bar{\omega}^2 S_D$$

$$= m\bar{\omega} S_V = m S_A \tag{7.24}$$

so that Eq. 7.23 is the equation of the maximum *absolute* acceleration since it corresponds to the maximum force in the member.

Noting the relationships between S_D, S_V and S_A, it is possible to construct a tripartite log plot of all three spectra to produce a combined spectrum

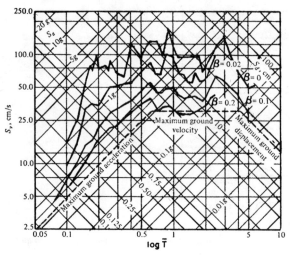

Fig. 7.9(a) Response spectra, for El Centro earthquake, 1940 (from Clough and Penzien[6])

such as that shown in Fig. 7.9(a)[6]. S_V is read vertically while S_A is read from lower right to upper left and S_D is read from lower left to upper right. Particularly interesting with respect to Fig. 7.9(a) is the dashed curve which represents the maximum *ground motions;* the difference between the structure response curves and the ground motions indicates the *amplification* of the ground motion through the structure and is very significant with respect to seismic design. Note that in the low period (high frequency) range, which corresponds to stiff systems, the maximum structure accelerations approach the maximum ground accelerations while the displacements remain small; on the other hand, in the high period (low frequency) range, which corresponds to flexible systems, the maximum structure displacements approach the maximum ground displacement while the accelerations remain small. In the intermediate range, which corresponds to many structural systems, considerable amplification of the ground motion is observed. For example at $\bar{T} = 0.5$ and $\beta = 0.02$, $S_A \simeq 0.9$ g while the maximum ground acceleration $z_g = 0.33$ g.

7.2.2 Inelastic spectrum

The procedure used to generate the response spectra may be extended to the system shown in Fig. 1.9. Since we are interested in maximum effects, we consider the maximum spring force as defined by Eq. 1.47 for $y = y_{max}$

$$Q_{max} = k y_Y = \bar{\omega}^2 m y_Y = \bar{\omega}^2 m y_{max}/\mu \qquad (7.25)$$

Using Newton's Law,

$$Q_{max} = m\ddot{x}_{max} \qquad (7.26)$$

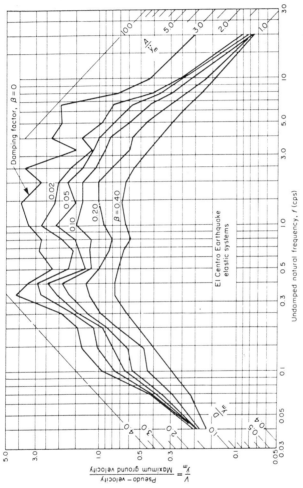

Fig. 7.9(b) Nondimensional response spectra for El Centro earthquake, 1940 (from Newmark[7])

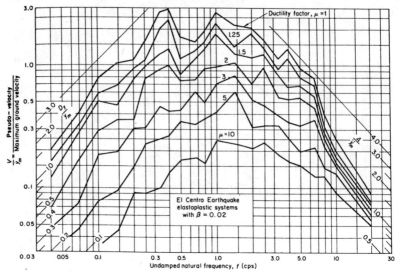

Fig. 7.9(c) Inelastic response spectrum for El Centro earthquake, 1940 (from Newmark[7])

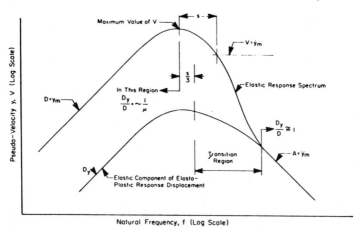

Fig. 7.9(d) Relation of elasto-plastic spectrum to elastic spectrum (from Newmark[7])

since $\dot{y}=0$ when $y=y_{max}$. Equating Eqs. 7.25 and 7.26, we have

$$\ddot{x}_{max}=(\bar{\omega}^2/\mu)y_{max} \tag{7.27}$$

Equation 1.49 may be solved for various values of $\bar{\omega}$ and β to obtain corresponding values of y_{max} after which the respective values of μ can be calculated from Eq. 1.48. Then the value of $\ddot{x}_{max}$ computed from Eq. 7.27 becomes the absolute spectral acceleration for a particular $\bar{\omega}$, β and μ, i.e.,

$$S_a=\ddot{x}_{max} \tag{7.28}$$

Equation 7.28 may be plotted in a similar manner to Fig. 7.8.

EARTHQUAKE EFFECTS ON STRUCTURES 141

Next we compute the pseudovelocity spectrum as before

$$S_V = (1/\bar{\omega})S_a \qquad (7.29)$$

similar to Eq. 7.6.
For the displacement, it is convenient to plot an "elastic" spectrum

$$S_{De} = y_{max}/\mu \qquad (7.30)$$

to accommodate the tripartite representation.
A nondimensionalized form of the elastic spectrum shown in Fig. 7.9(a) is shown in Fig. 7.9(b) and the corresponding inelastic spectrum for 2% damping is presented in Fig. 7.9(c)[7]. Because of the additional variable μ, it is possible to produce the spectrum only for a single value of β in each plot. From Fig. 7.9(c) it may be observed that in the low frequency range, the total displacement $\mu S_{De} (\mu D_y/y_m$ in figure) is about the same as the maximum displacement of an elastic system. In the high frequency range, the acceleration of the mass approaches the ground acceleration regardless of any inelastic action. However, in the intermediate range of interest, the inelastic action may reduce the amplifications and relieve the forces induced by the earthquake if the proper ductility can be provided. This is summarized in Fig. 7.9(d).

7.2.3 Design spectrum

In the design of actual structures, the response spectrum is a valuable resource and forms the basis for many current design techniques. However, many other factors may be included in the establishment of design criteria such as the probability of occurrence of a given ground motion, the damage that can be tolerated, the cost of repair or replacement as compared to the cost of providing additional resistance and the risk that might be assumed in a given situation. Many of these factors, including an estimate of inelastic behaviour, may be incorporated into a *design spectrum* which generally indicates smaller amplifications than the response spectrum. Additionally, a design spectrum, such as that shown in Fig. 7.10(a) or more legibly in Fig. 7.10(b)[7], may include the significant effects of many actual and even simulated earthquakes as opposed to the single event depicted by a response spectrum and is often "smoothed" to remove abrupt changes which have little chance of affecting an actual structure. While a response spectrum is the *effect* of a particular ground motion on various SDF oscillators, a design spectrum is a specification of the strength or deformation *capacity* required from a structure.

Insofar as the design spectrum is concerned, the level of damping chosen can affect the design forces (or required capacity) markedly. A smooth design spectrum is shown in Fig. 7.10(c) for various damping levels and is replotted in Fig. 7.10(d)[8] to show the design accelerations as functions of β

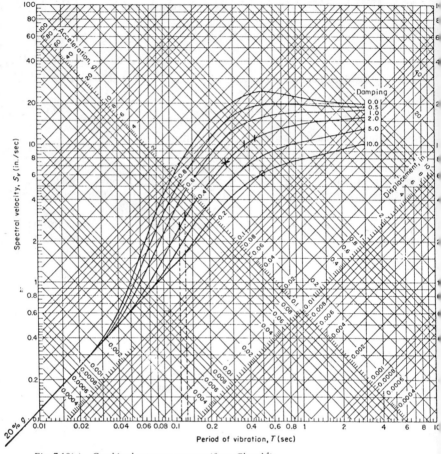

Fig. 7.10(a) Combined response spectra (from Clough[4])

for various periods. Field measurements on structures during earthquakes and full-scale simulation are helpful in establishing realistic damping factors.

In order to establish a level of confidence for an already designed structure to survive, the technique of *seismic assessment* is useful[9]. If a set of "credible excitations" is taken as those which could be caused by *any* ground motion which could be realistically expected to occur at the location of interest, then a given credible excitation is termed "critical" if it generates the largest response peak for a particular design variable. However the direct use of all critical excitations may lead to unrealistically severe results so that some selection and synthesis of credible ground motions may be desirable. This leads to the notion of "subcritical excitations" which generate response peaks similar to those produced by

EARTHQUAKE EFFECTS ON STRUCTURES 143

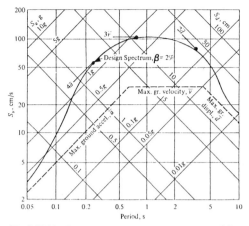

Fig. 7.10(b) Design response spectrum constructed from maximum ground motion characteristics (from Clough and Penzien[6])

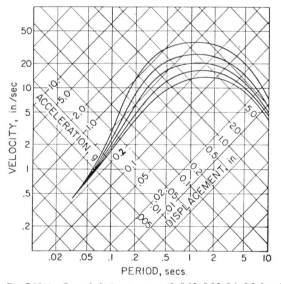

Fig. 7.10(c) Smooth design spectrum (0, 0.02, 0.05, 0.1, 0.2 fraction of critical damping) (from Housner and Jennings[8])

the critical excitation. The results of such a study indicate what ductility factors are required for the structure to survive the subcritical excitations. However, the technique as presently developed is confined to elastic behaviour and is thus approximate and limited.

In the same regard it should be noted that the response spectrum

144 EARTHQUAKE EFFECTS ON STRUCTURES

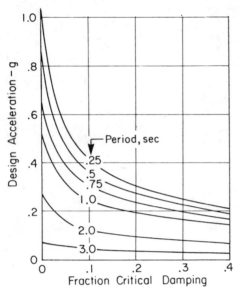

Fig. 7.10(d) Design spectrum as a function of damping (from Housner and Jennings[8])

technique as illustrated in the preceding section is not strictly applicable to MDF inelastic systems since superposition is invalid. For a more precise solution, a time history analysis may be required. Particularly, Anagnostopoulos, Haviland and Biggs[10] have found that axial forces in the exterior columns of multistory frames may be unconservatively predicted by modal analysis using inelastic spectra.

7.2.4 Forces in MDF systems

We may apply the response spectrum technique to MDF systems expressed in generalized coordinates as discussed in Section 4.7. In that section, the relative displacement of a MDF system was given by Eq. 4.62. By comparing Eq. 4.42 with Eqs. 7.18 to 7.20 and taking $\bar{p}_i$ in the form of Eq. 4.60, we find the maximum value of the generalized coordinate $q_i(t)$ to be

$$(q_i)_{\max} = \hat{p}_i/(\bar{m}_i \bar{\omega}_i) \cdot S_{V_i} \qquad (7.31)$$

where S_{V_i} is read from the spectrum opposite $\bar{\omega}_i$ and β_i. The results are best left in terms of S_{V_i} since it is easier to read than S_{D_i} or S_{A_i}.

Next we consider the elastic forces in the system. Equation 4.50 would take the form

$$[F]_{\max} = [K][\Phi]\{Q\}_{\max} \qquad (7.32)$$

where $\{Q\}_{max}$ is a vector of the $(q_i)_{max}$ computed by Eq. 7.31 from the spectrum ordinates. This expression may be overly conservative and indicates the major omission of spectral analysis. Namely, the maximum values of the quantities graphed in spectra i.e., displacements, velocities, accelerations, do not occur at the same time t for all modes in MDF systems. This so-called *phase difference* is lost in the response spectrum since only the maximum value of the Duhamel integral is recorded. In dealing with this situation, it is recommended that the forces and other responses associated with each mode be evaluated separately and then combined only as a final step, with the method of combination accounting for the phase difference. The square root of the sum of the squares of all contributing modal quantities (rss) is commonly used. In modal form, Eq. 7.32 becomes

$$\{F_i\}_{max} = [K]\{\varphi^i\}[q_i]_{max}$$

$$= [\hat{p}_i/(\bar{m}_i\bar{\omega}_i)][K]\{\varphi^i\}S_{V_i} \qquad (7.33)$$

It is common to express Eq. 7.33 in terms of the mass matrix. This is easily accomplished using Eqs. 4.3 which give, in view of Eqs. 4.4 and 4.5,

$$[K]\{\varphi^i\} = \bar{\omega}_i^2[M]\{\varphi^i\} \qquad (7.34)$$

whereupon

$$\{F_i\}_{max} = (\hat{p}_i\bar{\omega}_i/\bar{m}_i)[M]\{\varphi^i\}S_{V_i}$$

$$= (\hat{p}_i/\bar{m}_i)[M]\{\varphi^i\}S_{A_i} \qquad (7.35)$$

The evaluated set of elastic forces $\{F_i\}_{max}$ is conventionally applied to the MDF system, and the subsequent analysis for reactions, shears and bending moments is carried out on a mode-by-mode basis using static equilibrium. However, for continuous structures which have been discretized, such static procedures should be used cautiously since these elastic forces are essentially concentrated by the discretization. This is discussed further in Section 7.2.5.

Another quantity which is useful is the total force at the base of the structure which is equal to the sum of the elastic forces above the base. For the horizontal (u) component of an earthquake, this total force is called the *base shear* and is expressed by

$$Q_{o_i} = \sum^n F_{u_i}$$

$$= \lfloor I \rfloor \{F_{u_i}\}_{max}$$

$$= (\hat{p}_i\bar{\omega}_i/\bar{m}_i)\lfloor I \rfloor[M]\{\varphi^i\}S_{Vu_i} \qquad (7.36)$$

in which $\lfloor I \rfloor = \{I\}^T =$ the unit row vector.

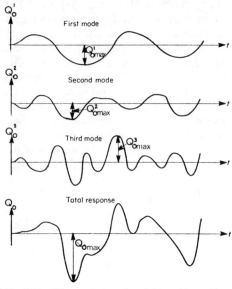

Fig. 7.11 *Time variation of modal base shears (from Clough[4])*

However, we note from Eq. 4.60 that the term

$$\lfloor I \rfloor [M]\{\varphi^i\} = \{\varphi^i\}^T[M]\{I\} = \hat{p}_i \quad (7.37)$$

so that

$$Q_{o_i} = W_i/g \cdot \bar{\omega}_i S_{Vu_i}$$

$$= W_i/g S_{Au_i} \quad (7.38)$$

in which

$$W_i = (\hat{p}_i^2/\bar{m}_i)g = \textit{effective weight} \text{ for mode } i \quad (7.39)$$

With the acceleration spectrum S_{Au} written as a fraction of the acceleration of gravity g, the base shear is clearly expressed as a fraction of the effective weight of the structure.

Again it should be noted that it is preferable to carry out the computations for all response quantities on a mode-by-mode basis and to combine only as a final step. To illustrate the phase difference, we refer to the plot of the base shear as a function of time for a 3DF system in Fig. 7.11. Note that the maximum for each mode is at different times and that the actual combined maximum occurs at a different time still. The rss combination is quite close for this case.

As an illustration we resume the analysis of the 2DF system shown in Fig. 4.2 and previously discussed at the end of Section 4.6. If we take $k/m = 500$, the circular frequencies are evaluated from Eqs. 4.54 as

$$\bar{\omega}_1 = 13.8 \text{ rad/s}; \quad \bar{\omega}_2 = 36.2 \text{ rad/s} \quad (7.40)$$

EARTHQUAKE EFFECTS ON STRUCTURES 147

Table 7.1

	S_D (in)	S_V (in/s)	S_A (%g)
$T_1 = 0.46$ s	0.80	11.5	0.38
$T_2 = 0.17$ s	0.12	4.4	0.40

from which the natural periods are

$$\bar{T}_1 = 2\pi/13.8 = 0.46 \text{ s}$$

$$\bar{T}_2 = 2\pi/36.2 = 0.17 \text{ s} \qquad (7.41)$$

Using Fig. 7.10(a) with $\beta = 2\%$, the spectral ordinates are summarized in Table 7.1.
Now we compute $\hat{p}_1$ from Eq. 4.60 as

$$\hat{p}_1 = \lfloor 1.000 \quad 1.618 \rfloor \begin{bmatrix} 1 & 0 \\ 0 & 1 \end{bmatrix} \begin{Bmatrix} 1 \\ 1 \end{Bmatrix} m = 2.62 \; m$$

$$\hat{p}_2 = \lfloor 1.000 \quad -0.618 \rfloor \begin{bmatrix} 1 & 0 \\ 0 & 1 \end{bmatrix} \begin{Bmatrix} 1 \\ 1 \end{Bmatrix} m = 0.38 \; m \qquad (7.42)$$

after which we evaluate $\{F_i\}_{\max}$ using Eq. 7.35:

$$\{F_i\}_{\max} = \left(\frac{2.62 \; m}{3.62 \; m}\right) \begin{bmatrix} 1 & 0 \\ 0 & 1 \end{bmatrix} m \begin{bmatrix} 1.000 \\ 1.618 \end{bmatrix} 0.38 \; g = \begin{Bmatrix} 0.27 \\ 0.44 \end{Bmatrix} mg$$

$$[F_i]_{\max} = \left(\frac{0.38 \; m}{1.38 \; m}\right) \begin{bmatrix} 1 & 0 \\ 0 & 1 \end{bmatrix} m \begin{Bmatrix} 1.000 \\ -0.618 \end{Bmatrix} 0.40 \; g = \begin{Bmatrix} 0.11 \\ -0.068 \end{Bmatrix} mg \qquad (7.43)$$

Next we find the effective weights from Eq. 7.39 as

$$W_1 = [(2.62 \; m)^2/3.62 \; m]g = 1.88 \; mg$$

$$W_2 = [(0.38 \; m)^2/1.38 \; m]g = 0.106 \; mg \qquad (7.44)$$

and finally the base shears from Eq. 7.38

$$Q_{o_1} = 1.88 \; mg/g \cdot 0.38 \; g = 0.71 \; mg$$

$$Q_{o_2} = 0.106 \; mg/g \cdot 0.40 \; g = 0.042 \; mg \qquad (7.45)$$

148 EARTHQUAKE EFFECTS ON STRUCTURES

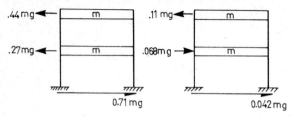

Fig. 7.12 Forces in 2DF system from response spectrum analysis

Table 7.2 SPECTRAL ACCELERATIONS AS A FUNCTION OF DUCTILITY RATIO

μ	$\bar{T}_1 = 0.46$ s $\bar{f}_1 = 2.17$ Hz	$\bar{T}_2 = 0.17$ s $\bar{f}_2 = 5.88$ Hz
1.0	0.55	0.70
1.25	0.50	0.58
1.5	0.35	0.50
2	0.25	0.40
3	0.13	0.28
5	0.08	0.16
10	0.06	0.10

with the rss value

$$(Q_0)_{rss} = [(0.71)^2 + (0.042)^2 mg]^{1/2} = 0.711 \, mg \qquad (7.46)$$

The final results are shown in Fig. 7.12. It may be noted that the second mode may significantly effect the second story shear even though it has little infleunce on $(Q_0)_{rss}$.

It is also of interest to determine the utilization of inelastic action to reduce the design forces, provided the assumed ductility can be provided. To do this, we refer to the response spectrum Fig. 7.9(c) and consider the 2DF system. In Table 7.2, the various spectral accelerations are given for ductility ratios 1, 1.25, 1.5, 2, 3, 5 and 10 with the base acceleration taken as 0.20 g so that a comparison with the values obtained from Fig. 7.10(a) may be made.

Obviously the spectra on Figs. 7.9(c) and 7.10(a) are not the same, even for $\mu = 1$, since the latter is a design spectrum. Nevertheless, the obvious reduction in spectral acceleration would indicate an approximately corresponding reduction in the elastic forces, effective weights and base shears as calculated in Eqs. 7.43 to 7.46. The superposed results must be interpreted as being only approximate as discussed in Section 7.2.3.

It is obvious from this example that a key factor in seismic resistant design, insofar as the structure is concerned, is the ductility that can be developed in conformity with the serviceability requirements of the structure.

EARTHQUAKE EFFECTS ON STRUCTURES   149

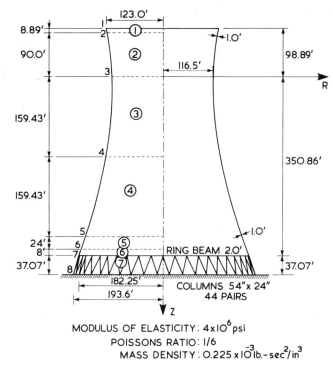

Fig. 7.13 *Hyperbolic cooling tower*

7.2.5 Seismic analysis of a hyperbolic cooling tower

To illustrate the application of the response spectrum procedure to a continuous system, the hyperbolic cooling tower shown in Fig. 7.13 is selected. This structure has been studied extensively in the literature[11,12] and is dimensionally representative of modern cooling towers.

With respect to the free vibration analysis for rotational shells, there are some significant differences as compared to lumped mass, shear type systems which are extensively discussed in the earthquake literature. Although the dynamic analysis of surface structures has not been so widely exposed, a number of effective procedures are available and the essentials have been introduced in Section 5.3.

For the hyperbolic cooling tower shown in Fig. 7.13, the results of the free vibration analysis are summarized in Table 7.3 for the $j=0$ (symmetric) and $j=1$ (antisymmetric) circumferential harmonics considering the first two longitudinal modes[12].

We observe that the natural frequencies are significantly affected when the system of supporting columns and the associated ring beam atop these columns are included in the model. This is accomplished through the use of

150 EARTHQUAKE EFFECTS ON STRUCTURES

Table 7.3 NATURAL FREQUENCIES $\bar{f}^{(j)}$, Hz

	Fixed base			Flexible base with ring beam	
j	Mode 1	Mode 2	j	Mode 1	Mode 2
0	6.558	10.117	0	5.873	9.487
1	2.709	5.752	1	2.326	3.834

a special 'open-type' element which is compatible with curved rotational shell elements[13].

In Fig. 7.14 corresponding mode shapes are shown following the notation of Fig. 5.1 and 5.2, except that z is measured from the base of the shell and is nondimensionalized by the total meridian length L_z. In Table 7.4, the maximum generalized coordinates for the first 3 modes of the $j = 1$ harmonic are given, based on the spectrum in Fig. 7.10 with $\beta = 0.02$[12].

We then compute the rss of the displacements by referring to Eqs. 5.47 to 5.51 with $j = 1$ and $n = 3$.

$$\{u^{(1)}\} = ([\Phi_u^{(1)}]^2 \{Q^{(1)}\}_{max}^2)^{1/2} \quad (a)$$

$$\{v^{(1)}\} = ([\Phi_v^{(1)}]^2 \{Q^{(1)}\}_{max}^2)^{1/2} \quad (b) \quad (7.47)$$

$$\{w^{(1)}\} = ([\Phi_w^{(1)}]^2 \{Q^{(1)}\}_{max}^2)^{1/2} \quad (c)$$

where the symbols $[\ ]^2$ and $\{\ \}^2$ means that each term of the matrix or vector is squared.

These displacements are shown in Figs. 7.15 to 7.17. Also shown, for purposes of comparison, are the combinations of the absolute values for the three modes (ABS) in each case. Particularly significant is the effect of the base flexibility on the deformations. From these figures, it may be surmised that the ductile performance of such a shell during an earthquake would depend primarily on the column-to-shell connection and the columns themselves, rather than on the shell.

The stress resultants and couples may also be calculated from the response spectrum results. As mentioned previously, the procedure often used for MDF systems is to compute a set of elastic forces, in this case harmonic by harmonic, using Eq. 7.35 and to then perform a static stress analysis; however, it has been shown that in curved shells serious inaccuracies may be introduced with this approach since the elastic forces $\{F_L^{(1)}\}$ will be line loads about the circumference.

A better procedure is to compute strains directly from the displacement

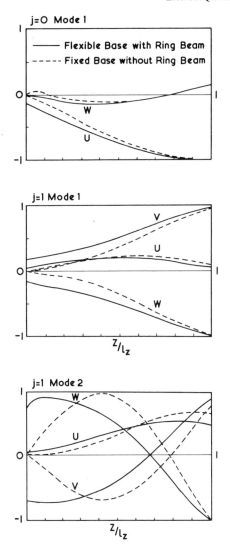

Fig. 7.14 Mode shapes for cooling tower (from Gould, et al.[12])

vectors, Eq. 7.47, and then to use the appropriate constitutive law to evaluate the stresses. Stress resultants and couples evaluated in this manner for the hyperbolic cooling tower are shown in Figs. 7.18 through 7.21[14]. These quantities are identified in Fig. 5.2. Particularly notable with respect to the influence of the flexible base conditions are the rapidly varying circumferential stress and the high localized bending near the base.

Table 7.4 GENERALISED COORDINATES FOR COOLING TOWER

Mode, i	$\bar{f}_i^{(1)}$	$\bar{\omega}_i^{(1)}$	$\bar{T}_i^{(1)}$	$\dfrac{p_i^{(1)}}{\bar{m}_i^{(1)}\bar{\omega}_i^{(1)}}$	$S_{V_i}^{(1)}$ (in/s)	$[q_i^{(1)}]_{max}$
Fixed base						
1	2.7087	17.0193	0.369	0.0909	10.0	0.909
2	5.7513	36.136	0.174	0.0219	4.7	0.103
3	9.1171	57.284	0.110	0.0126	2.4	0.030
Flexible base						
1	2.2966	14.4300	0.435	0.1078	11.1	1.197
2	3.8892	24.4365	0.257	0.0260	7.4	0.192
3	7.7279	48.5558	0.129	0.0035	2.9	0.010

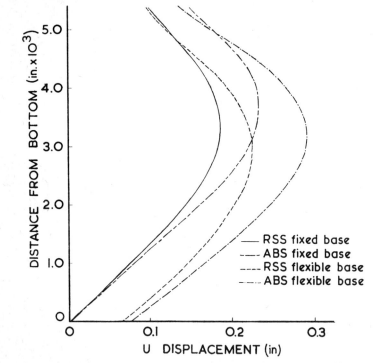

Fig. 7.15 Meridional displacement for cooling tower (from Gould, et al.[12])

7.3 TIME DOMAIN ANALYSIS

7.3.1 Direct integration of equations of motion

The modal approach used in the previous section leads to a Duhamel integral which may be evaluated at a particular time t to determine the instantaneous position of the system. For a MDF system, the procedure

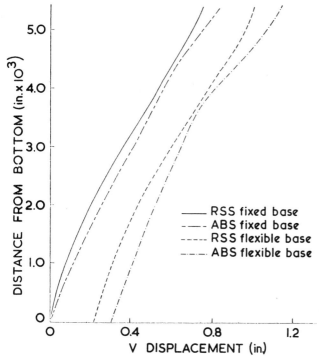

Fig. 7.16 Circumferential displacement for cooling tower (from Gould, et al[12])

requires that a free vibration analysis be completed prior to computing the time history response.

A somewhat different technique is to attempt to solve the system of Equations 4.1 directly. For this purpose, it is convenient to consider the incremental form

$$[M]\{\Delta \ddot{X}\} + [C]\{\Delta \dot{X}\} + [K]\{\Delta X\} = \{\Delta P\} \tag{7.48}$$

where the vectors $\{\Delta \ddot{X}\}$, $\{\Delta \dot{X}\}$ and $\{\Delta X\}$ represent the increment of the acceleration, velocity and displacement in each step Δt, and $\{\Delta P\}$ is the corresponding increment of the applied loading. Starting with a set of initial conditions at the time t_0, the response can be evaluated at subsequent times $t_1, t_2, \ldots, t_{j-1}, t_j, \ldots, t_{j+1} \ldots$ by using the displacement, velocity and acceleration at a time t_j (and possibly at $t_{j-1}, t_{j-2} \ldots$) to compute the corresponding quantities of t_{j+1}. The interested reader is referred to Dunham, et al[15] for a detailed discussion of direct integration procedures. Although a formal development of direct integration procedures is beyond the scope of this book, it is useful to consider the pros and cons of this approach.

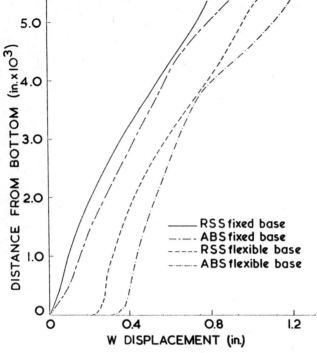

Fig. 7.17 Normal displacement for cooling tower (from Gould, et al.[12])

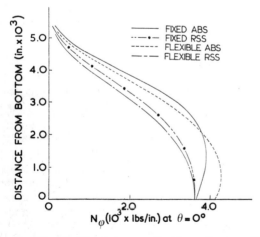

Fig. 7.18 Meridional stress resultant for cooling tower (from Gould, et al.[14])

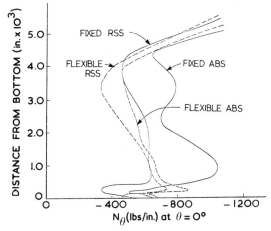

Fig. 7.19 Circumferential stress resultant for cooling tower (from Gould, et al.[14])

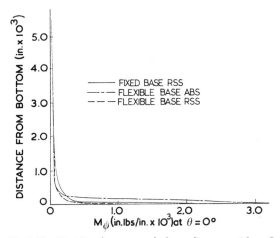

Fig. 7.20 Meridional stress couple for cooling tower (from Gould, et al.[14])

On the positive side, it is apparent that the eigenvalue problem may be avoided although in some instances it is very useful to have an estimate of the frequencies in order to select a proper time step. A more important advantage is in the analysis of nonlinear systems where the modal method is no longer applicable. It is obvious that the system matrices $[C]$ and $[K]$ can be updated at any time step reflecting geometric and/or material nonlinearities. Thus, the direct integration approach is considerably more general than the modal method.

On the negative side, the modal methods allow the analyst to consider the response in only those frequencies which are significant for the

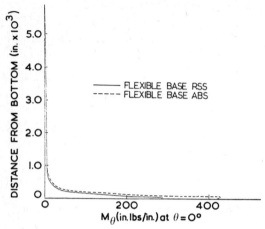

Fig. 7.21 Circumferential stress couple for cooling tower (from Gould, et al.[14])

particular excitation. In the case of earthquakes, the response is often adequately represented by the first two or three modes. Also, numerical stability may present a problem in some direct integration techniques.

In earthquake engineering, there are some cases in which a time history of the response, rather than only the spectral values, are needed. One such instance is in the analysis of secondary systems, as described in the next section.

7.3.2 Response of secondary systems

The modelling of complex structures for the purposes of dynamic analysis is a skill which often requires the consolidation of a large number of structural members into a reasonable number of primary components. This is necessary to keep the analysis to a manageable size and cost. The resulting consolidated model may be called the *primary* system and those components which are incorporated into *primary components* are termed *secondary* systems.

A simplistic but practical approach to this type of modelling is to consider the mass properties of the secondary components to be lumped at the node points of the primary system. It is evident that the shear-building model shown in Fig. 4.2 can accommodate the masses of all systems framing into each level. In some models, the stiffness properties of the secondary systems are ignored while in other models they may be included, at least approximately. The secondary systems are likely to add damping and this may also be included in the primary system model.

Examples of systems which might be classified as secondary for the purposes of the overall dynamic analysis are attached equipment, piping

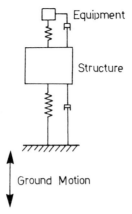

Fig. 7.22 SDF system with attached secondary system (from Stoykovitch[17])

and lightweight structures. Such systems have suffered a number of failures in earthquakes and some design guidelines for equipment have been compiled by Jordon[16]. It is evident that these items are definitely *not* secondary in importance insofar as the functional operation of the structure is concerned and that the effect of the earthquake on the secondary system itself is not predicted by the analysis of the primary system.

A straightforward approach is to use a two level analysis whereby:

(1) Time histories are generated for node points on the primary systems where secondary systems have been consolidated; and
(2) Secondary systems are regarded as SDF (or if necessary as MDF) systems subject to a base motion consisting of the time history at the point of attachment to the primary system as computed in (1).

A conceptual model of a SDF primary system with an attached secondary system is shown in Fig. 7.22[17].

Here, we generalize the concept to a MDF primary system. We assume that a modal (as opposed to a direct integration) analysis has been performed for an input base acceleration $\ddot{z}(t)$ and the system is described by Eq. 4.38 with the r.h.s. given by Eq. 4.61:

$$\ddot{q}_i + 2\beta\bar{\omega}_1\dot{q}_1 + \bar{\omega}_i^2 q_i = (\hat{p}_i/\bar{m}_i)\ddot{z}(t) \qquad (i=1,\ldots,n) \qquad (7.49)$$

In this case the negative sign on $\ddot{z}(t)$ is dropped as unnecessary.

Referring to Eqs. 4.42 and 4.60, the generalized coordinates expressed as a function of time may be written as

$$\{Q(t)\} = [\hat{P}][\hat{\Omega}]\{D(t)\} \qquad (7.50)$$

158   EARTHQUAKE EFFECTS ON STRUCTURES

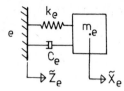

Fig. 7.23   Secondary system model

in which

$$\{Q(t)\} = \{q_1 q_2 \ldots q_n\}$$
$$[\hat{P}] = [\hat{p}_1/\bar{m}_1 \quad \hat{p}_2/\bar{m}_2 \ldots \hat{p}_n/\bar{m}_n]$$
$$[\hat{\Omega}] = [1/\bar{\omega}_1 1/\bar{\omega}_2 \ldots 1/\bar{\omega}_n]$$
(7.51)

and
$$\{D(t)\} = \{d_1 d_2 \ldots d_i \ldots d_n\}$$

where

$$d_i(t) = \int_0^t \ddot{z}(\tau) e^{-\beta_i \bar{\omega}_i(t-\tau)} \sin \bar{\omega}_i(t-\tau) d\tau \qquad (7.52)$$

We are also interested in the time history of the accelerations which are approximately

$$\{\ddot{Q}(t)\} = \lceil \bar{\Omega}^2 \rfloor \{Q(t)\}$$
$$= [\hat{P}][\bar{\Omega}]\{D(t)\} \qquad (7.53)$$

in which

$$\lceil \bar{\Omega} \rfloor = \lceil \bar{\omega}_1 \bar{\omega}_2 \ldots \bar{\omega}_n \rfloor \qquad (7.54)$$

From the preceding equations, the time histories of those mass points i at which the response of secondary systems are to be evaluated may be selected to complete part (1) of the analysis.

Now we consider a SDF secondary system attached at a point e subjected to the time history of response of the primary system $\tilde{z}_e(t)$ as shown in Fig. 7.23. With the mass and stiffness of the secondary system given by k_e and m_e, respectively, the circular frequency is $\bar{\omega}_e^2 = k_e/m_e$ and the damping is C_e or β_e. Just as in the earlier analysis of an SDF system subjected to base excitation (Section 1.5) it is convenient to work with the relative coordinate $\tilde{y}_e = \tilde{x}_e - \tilde{z}_e$. By analogy with Eq. 1.43, the equation of motion is

EARTHQUAKE EFFECTS ON STRUCTURES 159

$$\ddot{\tilde{y}}_e + 2\beta_e\bar{\omega}_e\dot{\tilde{y}}_e + \bar{\omega}_e^2\tilde{y}_e = -\ddot{\tilde{z}}_e \quad (7.55)$$

in which

$$\ddot{\tilde{z}}_e(t) = \ddot{z}(t) + \ddot{y}_e(t) \quad (7.56)$$

The relative acceleration at point i is given in terms of the generalized coordinates by Eq. 4.62 as

$$\ddot{y}_e(t) = \sum_{i=1}^{n} \varphi_e^i \ddot{q}_i(t) \quad (7.57)$$

where $n =$ the number of participating modes which may be equal to the total DOF of the main system, or which may possibly include selectively fewer modes. Then, substituting Eqs. 7.57 into Eq. 7.56

$$\ddot{\tilde{z}}_e = \ddot{z}(t) + \sum_{i=1}^{n} \varphi_e^i \ddot{q}_i \quad (7.58)$$

The solution to Eq. 7.55 follows from Eqs. 1.44 and 1.35(b) as

$$\tilde{y}_e(t) = 1/\bar{\omega}_e \cdot \tilde{d}_e(t) \quad (7.59)$$

where

$$\tilde{d}_e(t) = \int_0^t -\ddot{\tilde{z}}_e e^{-\beta_e \bar{\omega}_e(t-\tau)} \sin \bar{\omega}_e(t-\tau) d\tau \quad (7.60)$$

having the units of velocity.

Now, referring to Eqs. 7.18 to 7.20, it is apparent that the *maximum* values of Eq. 7.60 in the time range $\langle 0, t \rangle$ may be plotted as a function of $\bar{\omega}_e$ for various values of β_e to generate a response spectrum for point e. Such a spectrum is called a *floor response spectrum* and may be used to analyze a SDF secondary system, such as a piece of equipment, located at point e. Clearly, from Eq. 7.58, the floor response spectrum reflects the original ground motion plus the relative motion of the mass point e. Generally the acceleration spectrum $\bar{\omega}_e \tilde{d}_e$ is of most interest with respect to equipment.

Floor response spectra are informative as they may be quite different in shape as compared to the corresponding ground response spectrum. A representative spectrum is shown in Fig. 7.24[17] and clearly indicates that definite peaks are present in the floor response spectrum due to the filtering effect of the structure. Obviously, it is preferable to avoid frequency matching between the equipment and the peak points of the spectrum. If this is not possible, increasing the effective damping of the equipment may be helpful as indicated in Fig. 7.25.

An alternative to the floor response spectrum approach which is useful for estimating the maximum dynamic response of light attached equipment

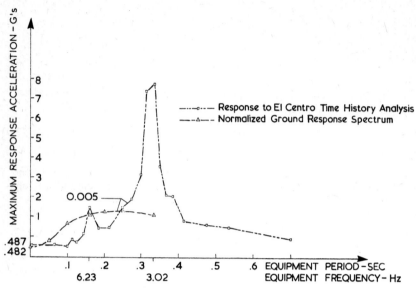

Fig. 7.24 Equipment response comparison (from Stoykovitch[17])

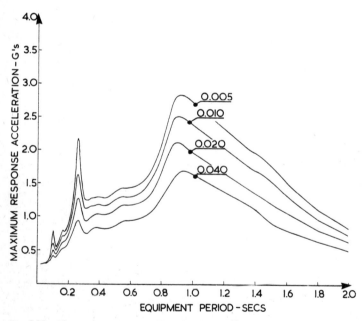

Fig. 7.25 Floor response spectra (from Stoykovitch[17])

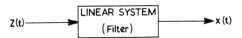

Fig. 7.26 *Linear system as a filter*

has been developed by Sackman and Kelley[27]. The charts summarizing their investigation can save significant computational effort since a time history analysis is avoided.

7.4 SYSTEMS REPRESENTATION

7.4.1 General

It is apparent that the motion observed at a given point on a structure has undergone several modifications since originating at the sources. Even if the base line, from the standpoint of seismic analysis of structures, is taken as an accelograph record at or near the surface, the structural response will likely be affected by the surrounding soil medium, the foundations and the remainder of the structure itself. We have touched on the latter effect in the discussion of floor responses in the preceding section but a more general approach is helpful, particularly for the conceptual understanding of the process.

7.4.2 Transfer functions

As a first step, we consider the linear filter shown in Fig. 7.26 where the linear system, however complex, is characterized as a filter such that for a unit input

$$z(t) = \delta(t) \tag{7.61}$$

the response will be

$$x(t) = h(t) \tag{7.62}$$

where $\delta(t)$ is the unit impulse function and $h(t)$ is the impulse response function, both introduced in Section 1.4. It is obvious that $h(t)$ characterizes the filter (or system); for the purposes of this development, it is helpful to call $h(t)$ the *transfer function* of the filter in the time domain.

When the input is more general, we may write the response as a convolution integral in the form of Eqs. 1.37 or 1.38

$$x(t) = z(t) * h(t) \tag{7.63}$$

where the standard notation (∗) has been introduced to indicate the convolution integral taken over a defined time range.

It is often convenient to characterize the system in the frequency domain.

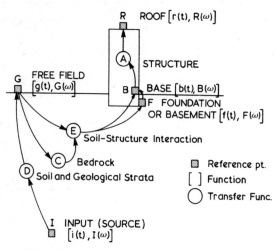

Fig. 7.27 *Transfer function representation of a source-soil-structure system*

We use the Fourier transforms

$$Z(\omega) = \int_{-\infty}^{\infty} z(t) e^{-i\omega t} dt \qquad (7.64)$$

and

$$H(\omega) = \int_{-\infty}^{\infty} h(t) e^{-i\omega t} dt \qquad (7.65)$$

in which $H(\omega)$ is called the transfer function in the frequency domain. Then the response to a general input is the simple product of the transforms[6a]

$$X(\omega) = Z(\omega) H(\omega) \qquad (7.66)$$

7.4.3 Source-soil-structure system

A conceptual model of the transmission of energy from a seismic source to key locations on a structure is shown in Fig. 7.27. This model is patterned after the systems representation of *Duke*[18]. It is conceivable that the same event could be measured simultaneously on accelographs located at the free field, G, at the base, B or the foundation, F, of the structure and on the roof, R; some field installations of this type are extant[19]. Between these reference

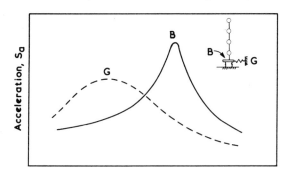

Fig. 7.28 Comparison of free field and base accelerations for soil-structure system (from Chu[21])

points, the medium is represented as a filter by an appropriate transfer function so that

$$G(\omega) = [D]I(\omega) \quad \text{(a)}$$
$$B(\omega) = [E]G(\omega) \quad \text{(b)} \quad (7.67)$$
$$R(\omega) = [A]B(\omega) \quad \text{(c)}$$

and the total response from source to roof is

$$R(\omega) = [A][E][D]I(\omega) \quad (7.68)$$

First we consider the filter [D], (soil and geological strata), although it might appear that the free field location, rather than the source, would be the place to start. Since a strong motion record may be available, some models require free field measurements to be 'deconvoluted' to the bed rock level whereupon the path would be directly to the foundation or base of the structure[20]. This is indicated by the alternative path through C. This is largely a geotechnical and geological problem and beyond our scope here.

The filter [E] (soil-structure interaction) is of considerable interest to structural engineers. The problem of determining [E] by comparison of measurements at B and G may be somewhat simplified if the transfer function is separated into separate horizontal, vertical and rotational components. A simplified model of comparative spectral accelerations is shown in Fig. 7.28[21] where the filtering effect of the soil medium and the shift in dominant frequency is apparent. A more elaborate model is shown in Fig. 7.29[21] where an additional reference point in the free field but at the level of F is established. Again the modification of the motion between I and F is apparent.

Finally the filter [A] (Structure) is represented by a dynamic model as described in Chapter 4. For example the maximum acceleration in mode i is

164 EARTHQUAKE EFFECTS ON STRUCTURES

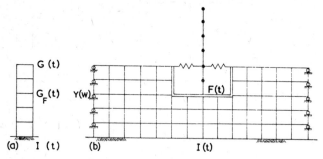

I(t) = Induced Rock Motion
$G_F(t)$ = Free Field Ground Motion at Foundation Level
G(t) = Free Field Ground Motion at Surface
F(t) = Basement Slab Motion

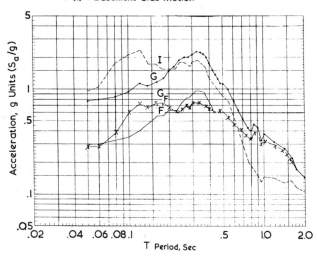

Fig. 7.29 Soil-structure interaction model; (a) free field model, (b) interaction model, (c) response spectra companion — Site A (from Chu[21])

obtained from Eq. 7.31, noting that $[\ddot{q}_i]_{max} \simeq \bar{\omega}_i^2 [q_i]_{max}$ as $[\ddot{q}_i]_{max} = (\hat{p}_i/\bar{m}_i) S_{A_i}$. Therefore the vector of maximum absolute acceleration responses is written from Eq. 4.29 as

$$\{\ddot{X}\}_{max} = rss([\Phi]\{\ddot{Q}\}_{max}) \quad (7.69)$$

where the detailed rss computation follows Eq. 7.47 in form. Then the horizontal transfer function may be approximated as a constant

$$A_H = \frac{[\ddot{x}_{roof}]_{max}}{[\ddot{x}_{base}]_{max}} \quad (7.70)$$

EARTHQUAKE EFFECTS ON STRUCTURES 165

Similar calculations will give the vertical and rotational components of A. There is a vast literature concerned with each class of transfer function discussed and, particularly with respect to soil-structure interaction, improved methods are rapidly emerging. A reasonably sophisticated finite element formulation is presented by Clough and Penzien[6b] while a simplified practical oriented representation is offered by Veletsos[22]. While the multitude of models and techniques may be difficult to reconcile, it is clear that the soil medium interacting with the lower part of the structure can considerably modify the response, as compared to that calculated if the free field motion is directly introduced at the base. The extent of this modification is dependent primarily on the characteristics of the soil strata. In many cases, the soil medium is a beneficial attendant to the structure as an energy absorber so that considerable design savings may accrue by the incorporation of soil-structure interaction.

Lastly, it should be mentioned that the quantitative treatment here has been based on a uniform ground motion over the entire site. Recent studies have shown that for structures which occupy a relatively large ground space, the travel time of the seismic wave and the resulting non-uniform motion may considerably alter the response of the structure[23].

7.5 NONDETERMINISTIC ANALYSIS

7.5.1 General

In comparison to the development of the deterministic methods of seismic analysis *viz a viz* the response spectrum method, corresponding stochastic procedures are not nearly as well developed. As mentioned previously, a design spectrum may include many factors which generalize the results of a single or even a series of response spectra. A design spectra recently published[24] incorporates randomness in the form of the standard deviation from the mean values. However, the direct application of random vibration techniques is hampered by limited strong-motion accelogram data[6c], and the relatively short duration of the event. Nevertheless the framework exists for the application of stochastic techniques to seismic analysis and such procedures may well become more popular as the available data base is increased.

7.5.2 Resonance excitation

It has been suggested that a seismic event can be adequately represented by a time history which consists of a portion of a stationary random process of finite duration, say 15–50 s. One form of the power spectrum of ground acceleration which is attributed to Tajimi[25] is

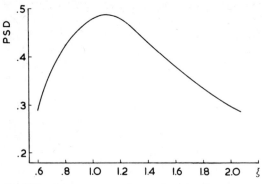

Fig. 7.30 Power spectrum of ground acceleration

$$\frac{\bar{\omega} S_{\ddot{z}}(\omega)}{\sigma_{\ddot{z}}^2} = \frac{4}{\pi} \frac{H}{1+4H^2} \cdot \frac{\xi[1+4H^2\xi^2]}{[(1-\xi^2)^2+4H^2\xi^2]} \qquad (7.71)$$

in which $\ddot{z}$ = ground acceleration, $\xi = \omega/\omega_g = \omega/(2\pi f_g)$, f_g = characteristic ground frequency of the order of 2 – 3 Hz, H = a nondimensional characteristic damping property of the ground, generally 0.6–0.7. The r.h.s. of Eq. 7.71 has been plotted in Fig. 7.30.

In terms of generalized coordinates, the generalized force is related to the ground acceleration through Eq. 4.61

$$\bar{p}_i = -\hat{p}_i \ddot{z}(t) \qquad (7.72)$$

so that

$$S_{\bar{p}_i} = \hat{p}_i S_{\ddot{z}(t)} \qquad (7.73)$$

whereupon the analysis follows the form presented in Chapter 3. Of course for a SDF system $\hat{p}_i^2 = m^2$.

We may compute the spectrum of the generalized displacement by substituting Eq. 7.73 into Eq. 5.13 as

$$S_{q_i} = (1/\bar{k}_i)^2 |H(\omega_i)|^2 S_{\bar{p}_i}(\omega_i) \qquad (7.74)$$

Also, the spectrum for a typical displacement y_k follows from Eq. 5.19 as

$$S_{y_k} = \sum_{i=1}^{n} S_{q_i}[\varphi^i(k)]^2 \qquad (7.75)$$

if modal cross-coupling is neglected.

If the response is dominated by resonance effects, the variance of the generalized coordinate is found from Eq. 3.13(a) as

$$\sigma_{q_i}^2 = (\pi/4\beta_i)(\bar{\omega}_i/\bar{k}_i^2)S_{p_i}(\bar{\omega}_i) \qquad (7.76a)$$

and the variance of the response is then

$$\sigma_{y_k}^2 = \sum_{i=1}^{n} \sigma_{q_i}^2 [\varphi^i(k)]^2 \qquad (7.76b)$$

Abu-Sitta and Davenport[26] applied this method to the analysis of a hyperbolic cooling tower which is similar to that discussed in Section 6.7 and shown in Fig. 6.18(a). They found that the response was dominated by the first circumferential mode ($j = 1$) and that an equivalent static loading which produces stresses close to those found from the dynamic analysis may be determined if $\bar{\omega}_1$ and f_g are known. The use of an equivalent static load may be helpful in preliminary considerations but such procedure is usually inadequate for detailed design computations.

REFERENCES

1. Housner, G. W. *Earthquake Engineering*, Ed. R. L. Weigel, Prentice-Hall, N. J., 1970, pp. 75–91
2. Bolt, B. A., *Earthquake Engineering*, Ed. R. L. Weigel, Prentice-Hall, N.J., 1970, pp. 1–20
3. Fagel, L. W. and Liu, S.-C., "Earthquake Interaction for Multistory Buildings," *J. Engng. Mech. Div., ASCE*, Vol. 98, No. EM4, Aug. 1972, pp. 929–945
4. Clough, R. W., *Earthquake Engineering*, Ed. R. L. Weigel, Prentice Hall, N.J., 1970, pp. 307–334
5. Rosenblueth, E. and Contreras, H., *Approximate Design for Multicomponent Earthquakes*," *J. Engng Mech. Div., ASCE*, Vol. 98, No. EM4, Aug. 1972, pp. 929–945
6. Clough, R. W. and Penzien, J., *Dynamics of Structures*, McGraw Hill, New York, 1975, pp. 537–543: (a) p. 468; (b) pp. 584–588; (c) p. 611
7. Newmark, N. M., *Earthquake Engineering*, Ed. R. L. Weigel, Prentice-Hall, N.J., 1970, pp. 403–424.
8. Housner, G. W. and Jennings, P. C., "Earthquake Design Criteria for Structures", EERL 77-06, California Inst. of Technology, Nov. 1977
9. Wang, P. C., Drenick, R. F. and Wang, W., "Seismic Assessment of High-Rise Buildings", *J. Engng Mech. Div. ASCE*, Vol. 104, No. EM2, April, 1978, pp. 441–456
10. Anagnostopoulos, S. A., Haviland, R. W. and Biggs, J. M., "Use of Inelastic Spectra in Aseismic Design", *J. Struct. Div. ASCE*, Vol. 104, No. S72, Jan. 1978, pp. 95–109
11. Abel, J. F., Cole, P. B. and Billington, D. B., "Maximum Seismic Response of Cooling Towers," *Res. Report* 73-2M-1, Dept. of Civil and Geological Engineering, Princeton University, 1973
12. Gould, P. L., Sen, S. K. and Suryoutomo, H., "Dynamic Analysis of Column-Supported Hyperboloidal Shells," *Int. J. Earthq. Engng Struct, Dyn.*, Vol. 2, 1974, pp. 269–279
13. Sen, S. K. and Gould, P. L., "Hyperboloidal Shells on Discrete Supports," (Tech. Note), *J. Struct. Div. ASCE*, Vol. 97, No. ST3, March, 1973, pp. 595–603
14. Gould, P. L., Suryoutomo, H. and Sen, S. K., "Stresses in Column-Supported Hyperboloidal Shells Subject to Seismic Loading," *Int. J. Earthq. Engng Struct. Dyn.*, Vol. 5, 1977, pp. 3–14

15. Dunham. R. S., Nickell, R. E. and Stickler, D. C., "Integration Operators for Transient Structure Response," *J. Computers and Structures*, Vol. 2, 1972, pp. 1–15
16. Jordan, C. H., "Seismic Restraint of Equipment in Buildings," *J. Struct. Div., ASCE*, Vol. 104, No. ST5, May 1978, pp. 829–839
17. Stoykovich, M., "Methods of Determining Floor Response Spectra," Proc. Sym. on Structural Design of Nuclear Power Plant Facilities, J. I. Abrams and J. O. Stevenson, Editors, U. of Pittsburgh School of Engineering Publication Series 9, Dec. 1952, pp. 114–166
18. Duke, G. M., "Earthquake Engineering Fundamentals," Continuous Education in Engineering and Science, University of California at Los Angeles, 1960
19. Hart, G. and Vasudevan, R., "Earthquake Design of Buildings: Damping," *J. Struct. Div. ASCE*, Vol. 101, No. ST1, Jan. 1975, pp. 11–30
20. Schnabel, P. B., Lysmer, J. and Seed, H. B., "Shake-A Computer Program for Earthquake Response Analysis of Horizontally Layered Sites," Earthquake Engineering Research Center Report EERC 72-12, Univ. of California, Berkeley, Dec. 1972
21. Chu, S. L., "Some Notes for Seismic Analysis on Nuclear Plant Facilities," *Proc. Sym. Structural Design of Nuclear Plant Facilities*, Ed. J. I. Abrams and J. O. Stevenson, U. of Pittsburgh, School of Engineering Publication Series 9, Dec. 1952, pp. 15–37
22. Veletsos, A. S., "Dynamics of Structure-Foundation Systems," in *Structural and Geotechnical Mechanics*, a Volume Honoring Nathan M. Newmark, Prentice Hall, N.J., 1977
23. Scanlan, R., "Seismic Wave Effects on Soil-Structure Interaction," *Int. J. Earthq. Engng Struct. Dyn.*, Vol. 4, 1976, pp. 379–388
24. Newmark, N. M., Blume, J. A. and Kapur, K. K., "Seismic Design Spectra for Nuclear Power Plants," *J. Power Div., ASCE*, Vol. 99, No. PO2, pp. 287–303
25. Tajimi, H., "A Statistical Method of Determining the Maximum Response of a Building Structure During an Earthquake," *Proc. 2nd World Conf. on Earthquake Engineering*, Vol. 2, Japan, 1960
26. Abu-Sitta, S. H. and Davenport, A. G., "Earthquake Design of Cooling Towers," *J. Struct. Div. ASCE*, Vol. 96, No. ST9, Sept. 1970, pp. 1889–1902
27. Sackman, J. L. and Kelley, J. M., 'Seismic Analysis of Internal Equipment and Components in Structures', *Engrg Structures*, Vol. 1, July 1979, pp. 179–190

8 Conclusion

In the preceding chapters, some state-of-the-art methods for treating wind and earthquake loadings on structures have been developed from a common mathematical and physical basis. The emerging procedures for each case are somewhat different in form however and it is instructive to review the major contributing factors. This question has been addressed recently by Davenport* and the succeeding remarks are based in part on his presentation.

First of all, both loadings have considerable fluctuating dynamic content but the duration of a severe wind storm may be several hours while a single earthquake shock is much briefer (Fig. 7.4(b)). Next, wind load fluctuations are conveniently measured from a static mean value as reflected in Fig. 6.20 while earthquakes do not appear to have such a bias and are therefore entirely dynamic. Also, the prevalent frequencies of dynamic wind loading are in the range of 1 cycle/min. and lower (Fig. 6.2) while the corresponding values for earthquake loading are in the 1–5 Hz range (Fig. 7.5).

These general observations are reflected in the response of structures to either type of lateral loading and also in the engineering approaches described in this book. The presence of a considerable nonfluctuating component in the total wind loading results in a largely static response of many structures to such a load. Moreover, the dominant frequencies present in an earthquake power spectrum are in the fundamental frequency range of many structures while the largest values shown in a wind spectrum usually fall below those of all but the most flexible structures. This means that the wind load response to the fluctuating component of the total loading is predominantly quasistatic while resonance effects may be very significant in the case of earthquake loading. In cases where inelastic behaviour is permitted, the longer duration and significant mean component characteristic of wind storms may produce more damage accumulation than that resulting from an earthquake.

When either of the lateral loadings are considered in the design criteria, the structure must naturally be provided with adequate strength to resist the induced forces. As such, the same resisting elements may be mobilized to counter either type of extreme lateral loading. An interesting basis of comparison as to which effect may be dominant, apart from seismological and climatic conditions, arises from the realization that wind loading is proportional to the *surface area* of the structure while the seismic loading is

---

* Davenport, A. G. "Comparison of Design Criteria for Wind and Seismic Loads," Session No. 47, ASCE Convention and Exposition, Chicago, Illinois, October 1978.

related directly to the *mass* or weight of the structure. Different classes of structures have quite different weight/frontal area ratios. Davenport has given values of 1000 lb/ft^2 for "heavy" structures such as tall buildings, 100 lb/ft^2 for suspension bridges and chimneys and 10 lb/ft^2 for "light" structures like signs and transmission lines. This indicates that earthquakes may be dominant for the heavier structures while wind could be more important for the lighter systems, all other things being equal.

Another factor to be considered is the spatial characteristics of the loading. It is apparent that the organization of wind as it affects structures is related to the size of a gust and the fundamental frequency of the structure. For a common height and frequency, the dynamic wind loading decreases as the overall size of the structure increases. This is recognised by the size reduction factor shown in Fig. 6.1(d). The physical size structure may also influence the loading induced into a structure by earthquake ground motion. The travel time of a shock wave across a large diameter structure and the corresponding interference together with changing material properties of the supporting medium may well mitigate the base accelerations as compared to those associated with the usual assumption of uniform ground motion. This will most likely be substantiated and quantified further in the future.

During the lifetime of the average structure, the occurrence of an extreme wind loading is far more probable than that of a large seismic loading. That is, the return period for a severe windstorm is considerably shorter than for a major earthquake. This is reflected in the more developed stochastic basis for wind loading considerations as opposed to the largely deterministic approach for earthquake loading. However, considerable strides have been made in characterizing seismic effects on a statistical basis and this effort will probably intensify.

The characteristics common to wind and earthquake effects on structures have been exploited to some extent in striving toward a unified design approach within the limitations imposed by the distinctly different natural phenomena and it is expected that more unification will be the trend for the future.

INDEX

Acceleration, building and human discomfort, 82
  spectrum, 130, 141, 146, 159
Accelerograph, 162
Acoustical radiation pressure, 116
Added mass ratio, 117
Aeolian, harp oscillations, 121–122
Aerodynamic, admittance function, 87, 91
  coefficients, 124
  damping force, 96
  instability, 121–122
  spoilers, 122
Air density, 118
Amplification of ground motion, 138
Amplitude, 3, 5, 6, 9, 42, 120
Artificial earthquake records, 141
Atmospheric turbulence, 75–80
Autocorrelation, function, 21–25, 28–29, 50, 54, 56, 75
Averaging times, wind speeds, 70–73
Axisymmetric, shells, 60
  structures, 59–67, 99

Background response, 91–93
Band width, 9, 23
Base, acceleration, 12, 47, 157
  disturbance 12, 47, 133
  motion, 12
  shear, 99, 145–148
Beams, prismatic, 48
Bluff body, 81
Body forces, 130
Boundaries of layered media, 131
Boundary layer, 68, 117

Cable, 53
  roofs, see Suspension roofs
Cantilever structures, 96, 97, 120
Cavity parameter, 117
Chimneys, tapered, 96–99, 121
Circular frequency, see Frequency
Circumferential strip, 99
Cladding, 83
Coefficient of variation, 16
Combined, random processes, 25–27
  spectrum, 137–138
Comfort criteria, 82
Complex frequency response, 6, 11, 30
Consistent mass matrix, 37
Continuous systems, 48–49
Convolution, 10–12, 28, 161
  See also *Duhamel's integral*

Cooling towers, see *Hyperbolic cooling towers*
Correlation, coefficient, 20
  length, 52
Co-spectra, 26
Coulomb friction, 13
Coupling of modes, 101
Covariance, 20–21, 51
Critical excitation, 142–143
Cross-correlation, 34, 64, 66, 78, 99, 101, 103, 115
  coefficient (coherence function), 26, 51–52, 56, 77, 103–107, 113
  function, 25, 34
Cross-spectral density (cross-spectra) 25–26, 34, 51, 64, 99
Cylindrical, body, 81, 99
  shell, 60
  tower, 99–115

d'Alembert's principle, 1
Damage, 141
Damped free vibration, 2, 4, 5
Dampers, see *Damping devices*
Damping, 1, 8, 9, 36, 44, 90–91, 141, 159
  aerodynamic, 5, 91, 96–97
  coefficient of critical, 4, 5, 44
  devices, 122
  matrix, 37
  evaluation of, 9
  parameters, 9
  ratio, 7
  structural, 121
  viscous, 1, 31
Deflection, 15, 54–55
Design spectrum, 141–144, 165
Diffraction, 131
Dirac delta function, 10, 25
Direct integration, 152–156
Discretization of continuous structures, 36
Displacement, 1, 13, 37, 61
  absolute, 12
  maximum, 13
  relative, 12, 47
  spectrum, 31, 57, 64, 118–119, 130, 141
  vector, 37
  yield, 13
Drag, 82, 122
  coefficient, 83, 96
Ductility, 13
  ratio, 14, 148
Duhamel's integral, 11, 44, 145, 152
  See also *Convolution*
Duration of earthquake, 131
Dynamic, pressure, 54, 81, 116–119

INDEX

Earthquake, classification systems, 128
  effects, 127–128
  elastic wave theory, 128–132
  the nature of, 127–132
Effective weight, 146–148
Eigenvalue, 37, 38, 155
Eigenvector, 37, 38
Elasto-Plastic model, 13–14
Elastic Wave model, 128–132
Elevated tank, 2
Energy absorbtion, 13
Ensemble, 19, 21
Equipment response, 156, 160
Ergodic random process, 21
Expectation, 16
Exposure, 70
Extreme value, 15
  distributions, 73–74

Fastest-mile wind speed, 70, 73
Fault displacement, 128
Filter, 161
Fisher-Tippet distributions, 74
Flexible base effects, 149–151
Floor response spectrum, 159
Flow, mean, 68
Fluctuations, 18, 68, 75, 116, 169
Flutter, 123–125
Focal depth, 127
Focus, 127
Forced vibration, 3
Forces, elastic, 45
Fourier integral, 22
  series, 9, 48, 59, 61, 62
  transform, 11, 24, 54, 162
Frequency, 6, 50, 57, 87, 91, 117, 146, 169–170
  characteristic, 166
  circular, 3–5, 24, 44, 146
  damped, 5
  domain, 24, 29, 51, 161
  forcing, 9
  matching, 9, 159
  natural, 3–5, 7, 9, 24
  predominant, 132–133
  reduced, 94
Friction, element, 13; wind velocity, 69, 77
Frontal region, 81, 99, 103, 107, 113
Fundamental period, see *Period*
Funnelling, 82

Gabled frame, 84, 86
Galloping, 122–123
Gaussian process, 19, 23
Generalized, coordinates, 42–47, 57, 63, 90, 144–145, 157, 166–167
  damping, 43
  displacement, 41, 166
  load, 43, 47–50, 56, 58, 166
  mass, 42, 48
  pressure spectrum, 56, 64, 89
  stiffness, 43, 48, 53
Geometrical coordinates, 37, 43–45
Gradient height, 68, 76
  velocity, 168
Ground motion, 14, 127–132, 162–165
Harmonic, excitation, 6, 31, 58
  motion, 5, 37
Helmholtz resonator, 59
Hourly wind speed, 70–74
Hyperbolic cooling tower, 59, 97, 100, 103, 104, 111–115, 149–151, 167

Impulse, 10–12
Incompressible wave propagation, 130
Incremental equations, 153
Indicial functions, 124
Inelastic, action, 148
  spectrum, 138–141
Inertia, 7, 8, 90
Inertial force, 1
Initial, conditions, 4, 6, 38, 40–42, 45
  displacement, 6, 40, 41, 44
  displacement vector, 46
  velocity, 6, 40, 41, 44
Input–output relationships, 161–165
Instability, see *Aerodynamic instability*
Intensity of earthquake, 128
Internal pressure, 83
Irrotational wave propogation, 130

Joint, acceptance, 51–52, 56
  probability density function, 19

Kinematic, law, 129
  viscosity, 81–82

Lateral response, 119–125
Lattice structure, 81
Layering effect, 131
Length scale, 77, 79–80
Lift, 82
  coefficient, 120
Loading, 15, 28, 99, 115
Load vector, 37, 47, 62
Local wind pressure, 83
Logarithmic decrement, 9, 121
Love waves, 131–132

Magnitude of earthquake, 128
Mass, 1, 2, 8, 36, 50, 63, 81, 121, 149, 158, 170
  matrix, 37, 62, 145
Masts, 121

INDEX 173

Mathematical models, 1–2
Mean, 16, 18, 50
 square, 16, 19, 22–23, 31, 51–54
Mechanical admittance function, 7, 30, 53, 91
 vibrators, 4
Membranes, 55, 58, 116, 118
Merchalli intensity scale, 128
Meridional ribs, 99–100
Modal coupling, 54, 57, 166
 damping coefficient, 44
 response, 47, 155
Mode shape, 37, 39, 41–42, 50, 55, 58, 60–61, 64, 118, 149–151
Moment, first, 18
 second, 18
Motion, equation(s) of, 1, 12, 36, 43
 perception of, 82
Multiple loads, 33–34
Multiple random variable (multivariate), 19–21
 multistory buildings, 133

Natural frequency, 3–5, 7, 37, 39, 46, 76, 106, 113, 121, 149–150
Narrow band filter, 23
Newton's law(s), 10, 138
Nondeterministic analysis of earthquakes, 165–167
Nonlinear systems, 155
Normal coordinates, 42
 mode, 42, 50

Orthogonal curvilinear coordinates, 59–60
Orthogonality, 39–40
Oscillator, 28–31
Overall structure, 83
Overdamped system, 5
Overturning moment, 99

Participation factor, 41, 47
Parts and portions of structure, 83
Peak, acceleration of earthquake, 131
 factor, 65, 76, 91–92, 104
 response, 65, 91, 113, 143
Period, 3, 16, 23, 135, 147
Periodic excitation, 7
 forcing function, 6–9
Phase difference, 145
Piping, 156
Plane structures, 55–59
Plates, 55, 58
Power, law formula, 68–70
 spectrum, 21–25
 spectral density, 23–24, 70, 76, 124

Pressure coefficients, 82–85, 95, 97, 100, 102, 103
 distribution, 52, 91–115
 spectrum, 53–54, 85–89, 106, 117–118
 waves, 131–132
Primary components, 156
Probability, 16–19, 73–74
 density function, 17, 19
 distribution function, 17
Pseudo velocity, 134, 141

Quadrature spectrum, 26, 99
Quasi-steady flow, 103
Quasistatic, 7–8, 32–33, 66, 101, 103, 113

Random, loading, 28, 33
 process, 15, 25
 variable, 15–16, 20
Rayleigh, damping, 44
 method, 48
Ritz method, 48
 waves, 131–132
Reactor housing, 59
Rectangular structures, 82–96
Recurrence interval, 132
Resonance, 7, 8, 9, 33, 57, 65, 91, 98, 103, 106, 113, 118, 120, 165–167
 factor, 114–115
Response, curve, 5
 spectrum, 31–32, 133–138, 141, 149–152, 165
Return period (mean recurrence intervals), 73, 74, 170
Reynold's number, 81–82, 96, 99, 100, 103, 119
Ribs, Cooling towers, see *Meridional ribs*
Richter scale, 128
Risk, 74
Root, mean-square, 19
 sum-square, 145, 150
Rotational shells, 60
Roughness factor, 91–92
 length, 70

Sample period, 15
Sampling, continuous, 15
 discrete, 15
Secondary systems, 156–161
Seismic, assessment, 142
 waves, 165
Self-excited oscillations, 119
Separation angle, 103, 104, 108
Shear waves, 130, 131, 132
Signal, 15, 23
Silo, 59

Single-degree-of-freedom (SDF) system(s), 1, 2, 4, 5, 9, 12, 13, 28, 31, 43, 48, 65, 120, 133, 134, 135, 157–159
Size reduction factor, 91, 92, 93
Slabs, 55
Slender flexible structures, 50–55
Soil, -structure interaction, 162–165
   geological strata, 163
Space-time correlation function, 50
Spatial, correlation of turbulence, 91
   separation, 52, 77
Spectrum, 75, 76, 77 (See various types of)
Spoiler devices, 122
Spring, 4
Stacks, see *Chimneys*
Standard deviation, 16, 65
Static response, 104
Stationary random, process, 21
   variable, 51
Statistical independence, 21
   parameters, 16, 21
Steady-state phase, 9
Stiffness, 8, 36
   matrix, 37, 62
Strain tensor, 129
Stress tensor, 129
Strouhal number, 82, 120
Structural dynamics, 1, 18
Suction, 100
Surface roughness, 81, 100, 101
Suspension bridges, 123–125
   roofs, 58, 115–119
Systems representation, 161–165

Tanks, 59, 97, 99
Terrain, 69, 71
Thickness, 99
Time, dependent loading, 11
   domain, 24, 152–156
   history, 133, 144, 157
Towers, 121
Transient phase, 9, 38

Transmission lines, 121–122
Transfer function, 29, 31, 161–165
Tripartite log plot, 137–138
Turbulence, 75, 79, 103
   intensity, 76
   wave length, 52, 83, 87, 97

Unit impulse function, 10, 12
Updrafts, see *Funnelling*

Variance, 16, 18, 20, 31, 51, 54, 57, 65–66, 75, 90, 167
   dimensionless, 16
Velocity, profile, 68–69, 100
   scaling law, 94
   spectrum, 134–135
Vertical strip, 99
Viscosity, 81–82
von Kármán Vortex Street, 119, 120
Vortex shedding, 81, 98, 119–123

Wake region, 81, 99, 103, 107
Wall, 118
Wave action, 15
White noise, 23, 25
Wind, effects, 68
   energy, 70
   loading, 15, 80–82, 113
   nature of, 68–70
   pressure, 80–119
   response of SDF system to, 31
   spectrum, 28, 31, 76–78, 169
   speed date, 70–75
   speed profiles, 68–70
   storm, 68
   turbulence pressure, 103, 116–117
   velocity profile, 69
   velocity records, 70–74

Zero crossing, rate of, 65
   plane displacement, 69

AUTHOR INDEX

Abel, J. F., 149, 167
Abu-Sitta, S. H., 33, 34, 58–59, 64–67, 103–113, 116–119, 126, 167, 168
ALCOA Conductor Products Co., 122, 126
Anagnostopolous, S. A., 144
ANSI, 70, 83–85, 94–97, 125
Ayre, R. S., 9, 48, 49

Basu, P. K., 115–116, 126
Beleveau, J. G., 124, 126
Bietry, J., 70, 125
Biggs, J. M., 144, 167
Billington, D. B., 115, 126, 149, 167
Blume, J. A., 165, 168
Bolt, B. A., 131–132, 167
Boyd, D. W., 70–71, 125
Bull, M. K., 117, 126

Chu, S. L., 163–164, 168
C.I.R.I.A., 122, 126
Clough, R. W., 6, 9, 19, 21, 24, 27, 29, 34, 39, 44, 48, 49, 51, 67, 135–138, 142–143, 146, 162, 165, 167
Cohen, E., 77, 122, 125
Cole, P. B., 149, 167
Contreras, H., 136, 167
Davenport, A. G., ix, 31, 33, 34, 51–55, 65, 67, 68–69, 91–95, 122, 125–126, 167–169
Drenick, R. F., 142, 167
Duke, G. M., 162, 168
Dunham, R. S., 153, 168

Elashkar, I. D., 58–59, 67, 116–119, 126

Fagel, L. W., 132–133, 167
Filliben, J. J., 74, 125
Fung, Y. C., 120–121, 123, 125–126

Gould, P. L., 60, 61–63, 67, 113, 115–116, 126, 149–156, 167
Gupta, K. K., 37, 49

Harris, R. J., 75, 77, 125
Hart, G., 162, 168
Hashish, M. G., 64–67, 103–113, 126
Haviland, R. W., 144, 167
Housner, G. W., 128, 141, 143–144, 167

Jacobsen, L. S., 9, 48, 49
Jennings, P. C., 141, 143–144, 167
Jordan, C. H., 157, 168

Kao, K. H., 86–87, 96, 98, 126
Kapur, K. K., 165, 168
Kelley, J. M., 159, 168
Kobayashi, H., 124, 126
Komatsu, S., 124, 126
Koppes, W. E., 82, 126
Kraus, H., 113, 126

Lin, W. H., 124, 126
Liu, S.-C., 132–133, 167
Lysmer, J., 163, 168

Macdonald, A. J., 78, 80, 119–120, 125

Newmark, N. M., 139–141, 165, 167–168
Nickell, R. E., 153, 168
Niemann, H. J., 100–102, 114–115, 126
Novak, M., 122, 126

Penzien, J., 6, 9, 19, 21, 24, 27, 29, 34, 39, 44, 48, 49, 51, 67, 136–138, 143, 162, 165, 167
Pröpper, H., 115, 126

Robson, J. D., 11, 16, 23, 24, 25, 27, 30, 33, 34
Rosenbluth, E., 136, 167

Sachs, P., 73, 125
Sackman, J. L., 159, 168
Sacre, C., 70, 125
Scanlan, R., 80, 82, 115, 124–126, 165, 168
Schnabel, P. B., 163, 168